# THE SIMPLIFIED STUDY GUIDE

## FOR THE

## GENERAL CLASS HAM RADIO

## LICENSE EXAMINATION

by

Rajiv C. Raju (K9RAJ)

**Platypus Global Media**

Published in the United States of America by:
**Platypus Global Media**
Registered with the U.S. ISBN agency
Contact Publisher at: ***pgmdirect@gmail.com***

Distribution by **Ingram Book Company**:
*www.ingrambook.com*

This edition first published September 2009

**ISBN-13: 978 0 9842212 0 2**
**ISBN-10: 0 9842212 0 4**

**© 2009 by Platypus Global Media**
All rights reserved worldwide.

Printed in the United States of America.

Some of the content in this book is from publicly available sources.

## Preface

The goal of this book is to distill the most essential knowledge necessary to pass the General Class License exam into a convenient and inexpensive package. The content of the book is based on publically available sources including the examination question pool, CFR Title 47 and Part 97 of the Amateur Radio Service. The author acknowledges the work of those who created the question pools for the Amateur Radio License exams. This book is valid till June 30, 2011. The book is not intended to be a substitute for more comprehensive books on the subject, such as the excellent book form the **ARRL**. It is only intended to help you pass the exam with a minimum investment in time. The reader is strongly encouraged to use the book from the **ARRL** to better understand the subject.

BLANK

# Acknowledgements

*This work is based on the publicly available question pool for Element 3 developed by the Question Pool Committee which represents the Volunteer Examiner Coordinators.*

BLANK

# Contents

| | | |
|---|---|---|
| I | RULES | 1 |
| II | PRACTICES | 7 |
| III | IMPORTANT FORMULAS | 11 |
| IV | COMMISSION RULES | 15 |
| V | OPERATING PROCEDURES | 33 |
| VI | PROPAGATION | 51 |
| VII | AMATEUR PRACTICES | 63 |
| VIII | ELECTRICAL PRINCIPLES | 79 |
| IX | CIRCUIT COMPONENTS | 89 |
| X | PRACTICAL CIRCUITS | 99 |
| XI | SIGNALS AND EMISSIONS | 109 |
| XII | ANTENNAS | 115 |
| XIII | SAFETY | 129 |

BLANK

# 1

# *RULES*

## The General Class License privileges.

The General Class license gives you privileges to operate in all amateur bands and modes, but you are not allowed to operate in the lowest frequency portions of some HF bands which are set aside for the Extra Class license holders.

Unless otherwise noted, the maximum power output is 1500 watts PEP. Although this is the maximum, the rules require that transmitter power must be the minimum necessary to carry out the desired communications. Geographical power restrictions apply to the 70 cm, 33 cm and 23 cm bands. At frequencies above 10m, the privileges are the same as those the Technician class license. Since you have already have passed the Technician class exam, you should be familiar with these privileges. Specific questions on the General Class test

SIMPLIFIED STUDY GUIDE FOR GENRAL CLASS...

will deal with the HF frequencies.

The General Class HF privileges are as follows:

**160 Meters**
1.800-2.000 MHz:CW,Phone,Image,RTTY/Data

**80 Meters**
3.525-3.600 MHz: CW, RTTY/Data
3.800-4.000 MHz: CW, Phone, Image

**60 Meters:**
Five Specific Channels using **USB voice only**
**50 Watts maximum ERP, 2.8 kHz bandwidth**
Channel Center / Amateur Tuning Frequency
5332 kHz / 5330.5 kHz
5348 kHz / 5346.5 kHz
5368 kHz / 5366.5 kHz
5373 kHz / 5371.5 kHz
5405 kHz (common US/UK) / 5403.5 kHz

**40 Meters**
7.025-7.125 MHz: CW, RTTY/Data
7.175-7.300 MHz: CW, Phone, Image

**30 Meters**
Maximum power**, 200 watts PEP**
10.100-10.150 MHz: **CW, RTTY/Data only**
**No phone or image permitted**

**20 Meters**
14.025-14.150 MHz: CW, RTTY/Data
14.225-14.350 MHz: CW, Phone, Image

**15 Meters**
21.025-21.200 MHz: CW, RTTY/Data
21.275-21.450 MHz: CW, Phone, Image

**12 Meters**
24.890-24.930 MHz: CW, RTTY/Data
24.930-24.990 MHz: CW, Phone, Image

**10 Meters**
28.000-28.300 MHz: CW, RTTY/Data
28.300-29.700 MHz: CW, Phone, Image

Note the special restrictions in the 60 meter and 30 meter bands. In the 60 meter and 30 meter bands amateur radio is not the primary service and must not cause interference with the primary services which are assigned these bands.

None of these bands are shared with the Citizens Radio Service

**Beacons** are only permitted a maximum transmit power of **100W PEP**. There can only be **one beacon per band** for any one **location.**

**Maximum Symbol Rates/Bandwidth for Digital Modes:**

| | |
|---|---|
| Below 10 m | -300 baud/ 1 kHz |
| 10 m | -1200 baud/ 1kHz |
| 6m, 2m | -19.6K baud/ 20 kHz |
| 1.25 m, 70 cm | -56K baud / 100 kHz |
| 33 cm and above | -no limits |

On the General class exam the questions dealing with symbol rates usually deal with RTTY.

There are restrictions on what hams can transmit. Obscene language and deceptive messages are not permitted. **Transmitting music is not permitted** except when it is **incidentally part of a space station transmission**. **Secret codes** are not allowed except to **send commands to space stations**. You may discuss selling your personal ham gear on the air as long as you do not make it a frequent or recurrent activity. You can not use ham radio for any other commercial purpose. Transmission of news is only permitted in an emergency when other means are not available.

Ham radio license examinations are administered by Volunteer Examiners (VE). To be a VE one must be accredited by a VEC, be 18 or older, have a general class or higher license and not have had any disciplinary action on their license by the FCC. You can only examine for license classes below your class of license. At least three Volunteer Examiners must be present to administer an exam. The VEs grade the exams and issue a **Certificate**

**of Successful Completion of Examination (CSCE)** which is good for 365 days. If you already have a call sign, you can start using general privileges as soon as you are issued a CSCE, but you have to say "temporary AG" after your call.

Third party communication occurs when information is sent to or from non-hams via hams. Third party traffic is only permitted via foreign amateurs if the country has an agreement permitting this with the United States, unless the traffic is related to emergencies or are messages related to ham radio/personal remarks. Individuals who have had disciplinary action on their ham radio license may not participate in third party traffic.

Repeaters simultaneously retransmit a received signal on another channel. The repeater may transmit on frequencies that are permitted by the repeater control operator's license. A repeater may participate in a voluntary coordination process to avoid interference with other repeaters. If an uncoordinated repeater interferes with a coordinated repeater, the uncoordinated repeater has the major responsibility to resolve the problem.

# 2

## PRACTICES

In SSB only one sideband is transmitted; the other sideband and carrier are suppressed. The main advantages of using single sideband as opposed to other voice modes on the HF amateur bands are that less bandwidth is used and there is high power efficiency. For this reason SSB is the commonly used voice mode on the High Frequency Amateur bands. By convention, phone communications on the bands at and above 20 meters use Upper Sideband (USB). Lower side band (LSB) is used on 160, 75 and 40 meter bands.

VOX allows "hands free" operation by using the user's voice to switch the microphone on. VOX settings include Anti-VOX, VOX Delay and VOX Sensitivity.

When using phone, the accepted procedure for breaking into a conversation is to say your call sign during a break between transmissions from the other stations.

"CQ DX" indicates the caller is looking for any station outside their own country.

If the frequency on which a net normally meets is in use just before the net begins, it is acceptable to ask the stations if the net may use the frequency, or move the net to a nearby clear frequency if necessary. If a net is about to begin on a frequency you and another station are using, then you should move to a different frequency as a courtesy to the net.

If you notice increasing interference from other activity on a frequency you are using, you should move your contact to another frequency. Ask if the frequency is in use, say your call sign, and listen for a response before calling CQ on phone.

The minimum frequency separation between signals is 150-500 Hz for CW, 3 kHz for SSB and 250-500 Hz for RTTY (170Hz shift).

A band plan is a voluntary guideline for band use beyond the FCC rules. A "DX Window" is an example of a band plan. This is a portion of the band that should not be used for contacts between stations within the 48 contiguous United States. When operating SSTV, RTTY and PSK you should follow generally accepted band plans.

When normal communications systems are not

available, an amateur station may use any means of radio communication at its disposal to provide essential communications when there is an immediate threat to the safety of human life or property. An amateur station is never prevented from using any means at its disposal to assist another station in distress. The location and nature of the emergency should be given to a station answering a distress transmission. Whatever frequency has the best chance of communicating the distress message should be used. Any mode may be used during a disaster.

Only a person holding an FCC issued amateur operator license may be the control operator of an amateur station transmitting in RACES to assist relief operations during a disaster. FCC may restrict normal frequency operations of amateur stations participating in RACES when the President's War Emergency Powers have been invoked.

Amateur volunteers who are formally enlisted to monitor the airwaves for rule violations belong to the Amateur Auxiliary. The objectives of the Amateur Auxiliary are to encourage amateur self-regulation and compliance with the rules.

SIMPLIFIED STUDY GUIDE FOR GENRAL CLASS...

# 3

## IMPORTANT FORMULAS

Ohm's Law:

**E=IR**
**I=E/R**
**R=E/I**

Decibels for power ratio:

**dB=10 log ($P_M$/$P_{REF}$)**

**Power ratio=anti log(dB/10)**

# SIMPLIFIED STUDY GUIDE FOR GENRAL CLASS...

Decibels for voltage ratio:

**dB=20 log ($V_M$/$V_{REF}$)**

**Voltage ratio=anti log(dB/20)**

Wavelength:

$$\lambda = c/f$$
$$f = c/\lambda$$
$$\lambda meters = 300/MHz$$

RMS for Sine wave:

**$V_{PEAK}$=$V_{PK\text{-}PK}$/2**

**$V_{RMS}$=(.707) x $V_{PEAK}$**

**PEP:**

**PEP=($V_{RMS}$)²/R**

Series Circuits:

$R_T = R_1 + R_2 + R_3$

$L_T = L_1 + L_2 + L_3$

$C_T = 1/(1/C_1 + 1/C_2 + 1/C_3)$

**Parallel circuits**

$R_T = 1/(1/R_1 + 1/R_2 + 1/R_3)$

$L_T = 1/(1/L_1 + 1/L_2 + 1/L_3)$

$C_T = C_1 + C_2 + C_3$

Transformers:

$E_S/E_P = N_S/N_P$

$Z_P = Z_S(N_P/N_S)^2$

# SIMPLIFIED STUDY GUIDE FOR GENRAL CLASS...

Reactance:

$$X_C = \frac{1}{2\pi f C}$$

$$X_L = 2\pi f L$$

The remainder of the book lists an edited version of the element 3 question pool. The wrong answer choices are edited out so the reader can more easily learn to recognize the correct answer choice. The recommended strategy is to read all of the questions three times, with the last reading being done as close to the exam as possible.

# COMMISSION RULES

SUBELEMENT G1 - COMMISSION'S RULES [5 Exam Questions - 5 Groups]

G1A - General class control operator frequency privileges; primary and secondary allocations

G1A01 (C) [97.301(d)]
On which of the following bands is a General Class license holder granted all amateur frequency privileges?
A. ---------
B. ---------
C. 160, 30, 17, 12, and 10 meters
D. ---------
~~

G1A02 (B) [97.305]
On which of the following bands is phone operation prohibited?
A. ----------
B. 30 meters
C. ----------
D. ----------
~~

# SIMPLIFIED STUDY GUIDE FOR GENRAL CLASS...

G1A03 (B) [97.305]
On which of the following bands is image transmission prohibited?
A. ---------
B. 30 meters
C. ---------
D. ---------
~~

G1A04 (D) [97.303(s)]
Which amateur band restricts communication to specific channels, using only USB voice, and prohibits all other modes, including CW and data?
A. ---------
B. ---------
C. ---------
D. 60 meters
~~

G1A05 (A) [97.301(d)]
Which of the following frequencies is in the General Class portion of the 40 meter band?
A. 7.250 MHz
B. ---------
C. ---------
D. ---------
~~

G1A06 (D) [97.301(d)]
Which of the following frequencies is in the 12 meter band?
A. ---------
B. ---------
C. ---------
D. 24.940 MHz
~~

G1A07 (C) [97.301(d)]
Which of the following frequencies is within the General class portion of the 75 meter phone band?
A. ---------
B. ---------
C. 3900 kHz
D. ---------
~~

G1A08 (C) [97.301(d)]
Which of the following frequencies is within the General Class portion of the 20 meter phone band?
A. ---------
B. ---------
C. 14305 kHz
D. ---------
~~

G1A09 (C) [97.301(d)]
Which of the following frequencies is within the General Class portion of the 80 meter band?
A. ---------
B. ---------
C. 3560 kHz
D. ---------
~~

G1A10 (C) [97.301(d)]
Which of the following frequencies is within the General Class portion of the 15 meter band?
A. ---------
B. ---------
C. 21300 kHz
D. ---------
~~

G1A11 (D) [97.301(d)]
Which of the following frequencies is available to a control operator holding a General Class license?
A. 28.020 MHz
B. 28.350 MHz
C. 28.550 MHz
D. All of these answers are correct

# SIMPLIFIED STUDY GUIDE FOR GENRAL CLASS...

~~

G1A12 (B) [97.301]
When a General Class licensee is not permitted to use the entire voice portion of a particular band, which portion of the voice segment is generally available to them?
A. ---------
B. The upper end
C. ---------
D. ---------
~~

G1A13 (D) [97.303]
Which amateur band is shared with the Citizens Radio Service?
A. ---------
B. ---------
C. ---------
D. None
~~

G1A14 (C) [97.303]
Which of the following applies when the FCC rules designate the amateur service as a secondary user and another service as a primary user on a band?
A. ---------
B. ---------
C. Amateur stations are allowed to use the frequency band only if they do not cause
   harmful interference to primary users
D. ---------
~~

G1A15 (D) [97.303]
What must you do if, when operating on either the 30 or 60 meter bands, a station in the primary service interferes with your contact?
A. ---------
B. ---------
C. ---------
D. Stop transmitting at once and/or move to a clear frequency
~~

G1A16 (A) [97.303(s)]
Which of the following operating restrictions applies to amateur radio stations as a secondary service in the 60 meter band?
A. They must not cause harmful interference to stations operating in other radio
   services
B. ---------
C. ---------
D. ---------
~~

G1B - Antenna structure limitations; good engineering and good amateur
practice; beacon operation; restricted operation; retransmitting radio signals

G1B01 (C) [97.15(a)]
What is the maximum height above ground to which an antenna structure may be erected without requiring notification to the FAA and registration with the FCC, provided it is not at or near a public-use airport?
A. ---------
B. ---------
C. 200 feet
D. ---------
~~

G1B02 (D) [97.203(b)]
With which of the following conditions must beacon stations comply?
A. ---------
B. ---------
C. ---------
D. There must be no more than one beacon signal in the same band from a single
   location
~~

# SIMPLIFIED STUDY GUIDE FOR GENRAL CLASS...

G1B03 (A) [97.1(a)(9)]
Which of the following is a purpose of a beacon station as identified in the FCC Rules?
A. Observation of propagation and reception, or other related activities
B. ---------
C. ---------
D. ---------
~~

G1B04 (A) [97.113(b)]
Which of the following must be true before an amateur station may provide news information to the media during a disaster?
A. The information must directly relate to the immediate safety of human life or
   protection of property and there is no other means of communication available
B. ---------
C. ---------
D. ---------
~~

G1B05 (D) [97.113(a)(4),(e)]
When may music be transmitted by an amateur station?
A. ---------
B. ---------
C. ---------
D. When it is an incidental part of a space shuttle or ISS retransmission
~~

G1B06 (B) [97.113(a)(4) and 97.207(f)]
When is an amateur station permitted to transmit secret codes?
A. ---------
B. To control a space station
C. ---------
D. ---------
~~

G1B07 (B) [97.113(a)(4)]
What are the restrictions on the use of abbreviations or procedural signals in
the amateur service?
A. ---------
B. They may be used if they do not obscure the meaning of a message
C. ---------
D. ---------
~~

G1B08 (D) [97.113(a)(4), 97.113(e)]
Which of the following is prohibited by the FCC Rules for amateur radio stations?
A. Transmission of music as the primary program material during a contact
B. The use of obscene or indecent words
C. Transmission of false or deceptive messages or signals
D. All of these answers are correct
~~

G1B09 (A) [97.113(a)(3)]
When may an amateur station transmit communications in which the licensee or control operator has a pecuniary (monetary) interest?
A. Only when other amateurs are being notified of the sale of apparatus normally
   used in an amateur station and such activity is not done on a regular basis
B. ---------
C. ---------
D. ---------
~~

G1B10 (C) [97.203(c)]
What is the power limit for beacon stations?
A. ---------
B. ---------
C. 100 watts PEP output
D. ---------
~~

# SIMPLIFIED STUDY GUIDE FOR GENRAL CLASS...

G1B11 (C) [97.101(a)]
How does the FCC require an amateur station to be operated in all respects not covered by the Part 97 rules?
A. ---------
B. ---------
C. In conformance with good engineering and good amateur practice
D. ---------
~~

G1B12 (A) [97.101(a)]
Who or what determines "good engineering and good amateur practice" that apply to operation of an amateur station in all respects not covered by the Part 97 rules?
A. The FCC
B. ---------
C. ---------
D. ---------
~~

G1B13 (A) [97.121(a)]
What restrictions may the FCC place on an amateur station that is causing interference to a broadcast receiver of good engineering design?
A. Restrict the amateur station operation to times other than 8 pm to 10:30 pm local
   time every day, as well as on Sundays from 10:30 am to 1 pm local time
B. ---------
C. ---------
D. ---------
~~


G1C - Transmitter power regulations; HF data emission standards


G1C01 (A) [97.313(c)(1)]
What is the maximum transmitting power an amateur station may use on 10.140 MHz?
A. 200 watts PEP output
B. ---------
C. ---------
D. ---------

G1C02 (C) [97.313(a),(b)]
What is the maximum transmitting power an amateur station may use on the 12 meter band?
A. ---------
B. ---------
C. 1500 watts PEP output
D. ---------~~

G1C03 (B) [97.313]
What is the maximum transmitting power a General class licensee may use when operating between 7025 and 7125 kHz?
A. ---------
B. 1500 watts PEP output
C. ---------
D. ---------
~~

G1C04 (A) [97.313]
What limitations, other than the 1500 watt PEP limit, are placed on transmitter power in the 14 MHz band?
A. Only the minimum power necessary to carry out the desired communications should
   be used
B. ---------
C. ---------
D. ---------
~~

G1C05 (C) [97.313]
What is the maximum transmitting power a station with a General Class control
operator may use on the 28 MHz band?
A. ---------
B. ---------
C. 1500 watts PEP output
D. ---------
~~

# SIMPLIFIED STUDY GUIDE FOR GENRAL CLASS...

G1C06 (D) [97.313(b)]
What is the maximum transmitting power an amateur station may use on 1825 kHz?
A. ---------
B. ---------
C. ---------
D. 1500 watts PEP output
~~

G1C07 (C) [97.303(s)]
Which of the following is a requirement when a station is transmitting on
the 60 meter band?
A. ---------
B. ---------
C. Transmissions must not exceed an effective radiated power of 50 Watts PEP
   referred to a dipole antenna
D. ---------
~~

G1C08 (D) [97.305(c) and 97.307(f)(3)]
What is the maximum symbol rate permitted for RTTY emissions transmitted on
frequency bands below 28 MHz?
A. ---------
B. ---------
C. ---------
D. 300 baud
~~

G1C09 (C) [97.305(c) and 97.307(f)(5)]
What is the maximum symbol rate permitted for packet emission transmissions on the 2 meter band?
A. ---------
B. ---------
C. 19.6 kilobaud
D. ---------
~~

G1C10 (C) [97.305(c) and 97.307(f)(4)]
What is the maximum symbol rate permitted for RTTY or data emission
transmissions on the 10 meter band?
A. ---------
B. ---------
C. 1200 baud
D. ---------
~~

G1C11 (B) [97.305(c) and 97.307(f)(5)]
What is the maximum symbol rate permitted for RTTY or data emission
transmissions on the 6 and 2 meter bands?
A. ---------
B. 19.6 kilobaud
C. ---------
D. ---------
~~

G1C12 (A) [97.305(c) and 97.307(f)(5)]
What is the maximum authorized bandwidth for RTTY, data or multiplexed
emissions using an unspecified digital code transmitted on the 6 and 2 meter
bands?
A. 20 kHz
B. ---------
C. ---------
D. ---------
~~

G1C13 (A) [97.303s]
What is the maximum bandwidth permitted by FCC rules for amateur radio
stations when operating on USB frequencies in the 60-meter band?
A. 2.8 kHz
B. ---------
C. ---------
D. ---------
~~

# SIMPLIFIED STUDY GUIDE FOR GENRAL CLASS...

G1D - Volunteer Examiners and Volunteer Examiner Coordinators; temporary identification

G1D01 (C) [97.119(f)(2)]
What is the proper way to identify when transmitting on General class frequencies if you have a CSCE for the required elements but your upgrade from Technician has not appeared in the ULS database?
A. ---------
B. ---------
C. Give your call sign followed by the words "temporary AG"
D. ---------
~~

G1D02 (C) [97.509(b)(3)(i)]
What license examinations may you administer when you are an accredited VE
holding a General Class operator license?
A. ---------
B. ---------
C. Technician
D. ---------
~~

G1D03 (C) [97.9(b)]
Which of the following band segments may you operate on if you are a Technician Class operator and have a CSCE for General Class privileges?
A. ---------
B. ---------
C. On any General Class band segment
D. ---------
~~

G1D04 (A) [97.509(a)(b)]
Which of the following are requirements for administering a Technician Class operator examination?
A. At Least three VEC-accredited General Class or higher VEs must be present
B. ---------
C. ---------
D. ---------
~~

26

G1D05 (D) [97.509(b)(3)(i)]
Which of the following is sufficient for you to be an administering VE for a Technician Class operator license examination?
A. ---------
B. ---------
C. ---------
D. A FCC General class or higher license and VEC accreditation
~~

G1D06 (A) [97.119(f)(2)]
When must you add the special identifier "AG" after your call sign if you are a Technician Class licensee and have a CSCE for General Class operator privileges?
A. Whenever you operate using General class frequency privileges
B. ---------
C. ---------
D. ---------
~~

G1D07 (B) [97.509(h)]
Who is responsible at a Volunteer Exam Session for determining the correctness of the answers on the exam?
A. ---------
B. The administering VEs
C. ---------
D. ---------
~~

G1D08 (B) [97.509(i)]
What document must be issued to a person that passes an exam element?
A. ---------
B. CSCE
C. ---------
D. ---------
~~

# SIMPLIFIED STUDY GUIDE FOR GENRAL CLASS...

G1D09 (C) [97.3(a)(15)]
How long is a Certificate of Successful Completion of Examination(CSCE)valid for exam element credit?
A. ---------
B. ---------
C. 365 days
D. ---------
~~

G1D10 (B) [97.509(b)(2)]
What is the minimum age that one must be to qualify as an accredited Volunteer Examiner?
A. ---------
B. 18 years
C. ---------
D. ---------
~~

G1D11 (B) [97.509 (b)(3)]
What criteria must be met for a non U.S. citizen to be an accredited Volunteer Examiner?
A. ---------
B. The person must hold a U.S. amateur radio license of General class or above
C. ---------
D. ---------
~~

G1D12 (C) [97.509(b)(1)]
Volunteer Examiners are accredited by what organization?
A. ---------
B. ---------
C. A Volunteer Examiner Coordinator
D. ---------
~~

G1D13 (D) [97.509]
When may you participate as a VE in administering an amateur radio license examination?
A. ---------
B. ---------
C. ---------
D. Once you have been granted your General class license and received your VEC
　accreditation
~~

G1E - Control categories; repeater regulations; harmful interference; third party rules; ITU regions

G1E01 (A) [97.115(b)(2)]
Which of the following would disqualify a third party from participating in stating a message over an amateur station?
A. The third party is a person previously licensed in the amateur service whose
　license had been revoked
B. ---------
C. ---------
D. ---------
~~

G1E02 (D) [97.205(a)]
When may a 10 meter repeater retransmit the 2 meter signal from a station having a Technician Class control operator?
A. ---------
B. ---------
C. ---------
D. Only if the 10 meter control operator holds at least a General class license
~~

G1E03 (A) [97.3(a)(39)]
What kind of amateur station simultaneously retransmits the signals of other stations on another channel?
A. Repeater Station
B. ---------
C. ---------
D. ---------
~~

# SIMPLIFIED STUDY GUIDE FOR GENRAL CLASS...

G1E04 (D) [97.13(b),97.311(b), 97.303]
Which of the following conditions require an amateur radio station to take specific steps to avoid harmful interference to other users or facilities?
A. When operating within one mile of an FCC Monitoring Station
B. When using a band where the amateur service is secondary
C. When a station is transmitting spread spectrum emissions
D. All of these answers are correct
~~

G1E05 (C) [97.115(a)(2), 97.117]
What types of messages for a third party in another country may be transmitted by an amateur station?
A. ---------
B. ---------
C. Only messages relating to amateur radio or remarks of a personal character, or
   messages relating to emergencies or disaster relief
D. ---------
~~

G1E06 (A) [97.205(c)]
Which of the following applies in the event of interference between a coordinated repeater and an uncoordinated repeater?
A. The licensee of the non-coordinated repeater has primary responsibility to resolve
   the interference
B. ---------
C. ---------
D. ---------
~~

G1E07 (C) [97.115(a)(2)]
With which of the following is third-party traffic prohibited, except for messages directly involving emergencies or disaster relief communications?
A. ---------
B. ---------
C. Any country other than the United States, unless there is a third-party agreement
   in effect with that country
D. ---------
~~

G1E08 (B) [97.115(a)(b)]
Which of the following is a requirement for a non-licensed person to communicate with a foreign amateur radio station from a US amateur station at which a licensed control operator is present?
A. ---------
B. The foreign amateur station must be in a country with which the United States has
   a third party agreement
C. ---------
D. ---------~~

G1E09 (C) [97.119(b)(2)]
What language must you use when identifying your station if you are using a language other than English in making a contact?
A. ---------
B. ---------
C. English
D. ---------
~~

G1E10 (D) [97.115(a)(2)]
Which of the following is a permissible third party communication during routine amateur radio operations?
A. ---------
B. ---------
C. ---------
D. Sending a message to a third party through a foreign station, as long as that
   person is a licensed amateur radio operator
~~

# OPERATING PROCEDURES

SUBELEMENT G2 - OPERATING PROCEDURES [6 Exam Questions - 6 Groups]

G2A Phone operating procedures; USB/LSB utilization conventions; procedural signals; breaking into a QSO in progress; VOX operation

G2A01 (A)
Which sideband is most commonly used for phone communications on the bands above 20 meters?
A. Upper Sideband
B. ---------
C. ---------
D. ---------
~~

# SIMPLIFIED STUDY GUIDE FOR GENRAL CLASS...

G2A02 (B)
Which sideband is commonly used on the 160, 75, and 40 meter bands?
A. ---------
B. Lower Sideband
C. ---------
D. ---------
~~

G2A03 (A)
Which sideband is commonly used in the VHF and UHF bands?
A. Upper Sideband
B. ---------
C. ---------
D. ---------
~~

G2A04 (A)
Which mode is most commonly used for voice communications on the 17 and 12 meter bands?
A. Upper Sideband
B. ---------
C. ---------
D. ---------
~~

G2A05 (C)
Which mode of voice communication is most commonly used on the High Frequency Amateur bands?
A. ---------
B. ---------
C. SSB
D. ---------~~

G2A06 (B)
Which of the following is an advantage when using single sideband as compared to other voice modes on the HF amateur bands?
A. ---------
B. Less bandwidth used and high power efficiency
C. ---------
D. ---------
~~

G2A07 (B)
Which of the following statements is true of the single sideband (SSB) voice mode?
A. ---------
B. Only one sideband is transmitted; the other sideband and carrier are suppressed
C. ---------
D. ---------
~~

G2A08 (A)
Which of the following statements is true of single sideband (SSB) voice mode?
A. It is a form of amplitude modulation in which one sideband and the carrier are
   suppressed
B. ---------
C. ---------
D. ---------
~~

G2A09 (D)
Why do most amateur stations use lower sideband on the 160, 75 and 40 meter bands?
A. ---------
B. ---------
C. ---------
D. Current amateur practice is to use lower sideband on these frequency bands
~~

G2A10 (B)
Which of the following statements is true of VOX operation?
A. ---------
B. VOX allows "hands free" operation
C. ---------
D. ---------
~~

# SIMPLIFIED STUDY GUIDE FOR GENRAL CLASS...

G2A11 (D)
Which of the following user adjustable controls are usually associated with VOX circuitry?
A. Anti-VOX
B. VOX Delay
C. VOX Sensitivity
D. All of these choices are correct
~~

G2A12 (B)
What is the recommended way to break into a conversation when using phone?
A. ---------
B. Say your call sign during a break between transmissions from the other stations
C. ---------
D. ---------
~~

G2A13 (C)
What does the expression "CQ DX" usually indicate?
A. ---------
B. ---------
C. The caller is looking for any station outside their own country
D. ---------
~~


G2B - Operating courtesy; band plans


G2B01 (C)
What action should be taken if the frequency on which a net normally meets is in use just before the net begins?
A. ---------
B. ---------
C. Ask the stations if the net may use the frequency, or move the net to a nearby
   clear frequency if necessary
D. ---------
~~

G2B02 (A)
What should be done if a net is about to begin on a frequency you and another station are using?
A. Move to a different frequency as a courtesy to the net
B. ---------
C. ---------
D. ---------
~~

G2B03 (C)
What should you do if you notice increasing interference from other activity on a frequency you are using?
A. ---------
B. ---------
C. Move your contact to another frequency
D. ---------
~~

G2B04 (B)
What minimum frequency separation between CW signals should be allowed to minimize interference?
A. ---------
B. 150 to 500 Hz
C. ---------
D. ---------
~~

G2B05 (B)
What minimum frequency separation between SSB signals should be allowed to minimize interference?
A. ---------
B. Approximately 3 kHz
C. ---------
D. ---------
~~

G2B06 (B)
What minimum frequency separation between 170 Hz shift RTTY signals should be allowed to minimize interference?
A. ---------
B. 250 to 500 Hz
C. ---------
D. ---------

# SIMPLIFIED STUDY GUIDE FOR GENRAL CLASS...

~~

G2B07 (A)
What is a band plan?
A. A voluntary guideline for band use beyond the divisions established by the FCC
B. ---------
C. ---------
D. ---------
~~

G2B08 (A)
What is the "DX window" in a voluntary band plan?
A. A portion of the band that should not be used for contacts between stations within
   the 48 contiguous United States
B. ---------
C. ---------
D. ---------
~~

G2B09 (D)
What should you do to comply with good amateur practice when choosing a frequency for Slow-Scan TV (SSTV) operation?
A. ---------
B. ---------
C. ---------
D. Follow generally accepted band plans for SSTV operation
~~

G2B10 (D)
What should you do to comply with good amateur practice when choosing a frequency for radio-teletype (RTTY) operation?
A. ---------
B. ---------
C. ---------
D. Follow generally accepted band plans for RTTY operation
~~

G2B11 (D)
What should you do to comply with good amateur practice when choosing a frequency for HF PSK operation?
A. ---------
B. ---------
C. ---------
D. Follow generally accepted band plans for PSK operation
~~

G2B12 (A)
What is a practical way to avoid harmful interference when selecting a frequency to call CQ using phone?
A. Ask if the frequency is in use, say your callsign, and listen for a response
B. ---------
C. ---------
D. ---------
~~

G2B13 (C)
What is a practical way to avoid harmful interference when calling CQ using Morse code or CW?
A. ---------
B. ---------
C. Send "QRL? de" followed by your callsign and listen for a response
D. ---------
~~


G2C - Emergencies, including drills and emergency communications


G2C01 (C) [97.403]
When normal communications systems are not available, what means may an amateur station use to provide essential communications when there is an immediate threat to the safety of human life or the protection of property?
A. ---------
B. ---------
C. Any means of radiocommunication at its disposal
D. ---------
~~

# SIMPLIFIED STUDY GUIDE FOR GENRAL CLASS...

G2C02 (A) [97.407(a)]
Who may be the control operator of an amateur station transmitting in RACES to assist relief operations during a disaster?
A. Only a person holding an FCC issued amateur operator license
B. ---------
C. ---------
D. ---------
~~

G2C03 (D) [97.407(b)]
When may the FCC restrict normal frequency operations of amateur stations participating in RACES?
A. ---------
B. ---------
C. ---------
D. When the President's War Emergency Powers have been invoked
~~

G2C04 (C) [97.405(b)]
When is an amateur station prevented from using any means at its disposal to assist another station in distress?
A. ---------
B. ---------
C. Never
D. ---------
~~

G2C05 (B) [97.403]
What type of transmission would a control operator be making when transmitting out of the amateur band without station identification during a life threatening emergency?
A. ---------
B. An unidentified transmission
C. ---------
D. ---------
~~

G2C07 (B)
What is the first thing you should do if you are communicating with another amateur station and hear a station in distress break in?
A. ---------
B. Acknowledge the station in distress and determine what assistance may be needed
C. ---------
D. ---------
~~

G2C08 (C) [97.405(b)]
When are you prohibited from helping a station in distress?
A. ---------
B. ---------
C. You are never prohibited from helping any station in distress
D. ---------
~~

G2C09 (B) [97.111(a)(2)]
What type of transmissions may an amateur station make during a disaster?
A. ---------
B. Transmissions necessary to meet essential communications needs and
   to facilitate relief actions
C. ---------
D. ---------
~~

G2C10 (C)
Which emission mode must be used to obtain assistance during a disaster?
A. ---------
B. ---------
C. Any mode
D. ---------
~~

# SIMPLIFIED STUDY GUIDE FOR GENRAL CLASS...

G2C11 (B)
What information should be given to a station answering a distress transmission?
A. ---------
B. The location and nature of the emergency
C. ---------
D. ---------
~~

G2C12 (A)
What frequency should be used to send a distress call?
A. Whatever frequency has the best chance of communicating the distress
   message
B. ---------
C. ---------
D. ---------
~~

G2D - Amateur auxiliary; minimizing Interference; HF operations

G2D01 (A)
What is the Amateur Auxiliary to the FCC?
A. Amateur volunteers who are formally enlisted to monitor the airwaves for rules
   violations
B. ---------
C. ---------
D. ---------
~~

G2D02 (B)
What are the objectives of the Amateur Auxiliary?
A. ---------
B. To encourage amateur self-regulation and compliance with the rules
C. ---------
D. ---------
~~

G2D03 (B)
What skills learned during "Fox Hunts" are of help to the Amateur Auxiliary?
A. ---------
B. Direction-finding skills used to locate stations violating FCC Rules
C. ---------
D. ---------
~~

G2D04 (B)
What is an azimuthal projection map?
A. ---------
B. A world map projection centered on a particular location
C. ---------
D. ---------
~~

G2D05 (A)
What is the most useful type of map to use when orienting a directional HF
antenna toward a distant station?
A. Azimuthal projection
B. ---------
C. ---------
D. ---------
~~

G2D06 (C)
How is a directional antenna pointed when making a "long-path" contact with another station?
A. ---------
B. ---------
C. 180 degrees from its short-path heading
D. ---------
~~

G2D07 (B) [97.103b]
Which of the following information must a licensee retain as part of their station records?
A. ---------
B. Antenna gain calculations or manufacturer's data for antennas used on 60 meters
C. ---------
D. ---------

# SIMPLIFIED STUDY GUIDE FOR GENRAL CLASS...

~~

G2D08 (D)
Why do many amateurs keep a log even though the FCC doesn't require it?
A. ---------
B. ---------
C. ---------
D. To help with a reply if the FCC requests information on who was control operator
   of your station at a given date and time
~~

G2D09 (D)
What information is traditionally contained in a station log?
A. Date and time of contact
B. Band and/or frequency of the contact
C. Call sign of station contacted and the signal report given
D. All of these choices are correct
~~

G2D10 (B)
What is QRP operation?
A. ---------
B. Low power transmit operation, typically about 5 watts
C. ---------
D. ---------
~~

G2D11 (C)
Which HF antenna would be the best to use for minimizing interference?
A. ---------
B. ---------
C. A unidirectional antenna
D. ---------
~~

G2D12 (A) [97.303s]
Which of the following is required by the FCC rules when operating in the 60 meter band?
A. If you are using other than a dipole antenna, you must keep a record of the gain
   of your antenna
B. ---------
C. ---------
D. ---------
~~

G2E - Digital operating: procedures, procedural signals and common abbreviations

G2E01 (D)
Which mode should be selected when using a SSB transmitter with an Audio Frequency Shift Keying (AFSK) RTTY signal?
A. ---------
B. ---------
C. ---------
D. LSB
~~

G2E02 (A)
How many data bits are sent in a single PSK31 character?
A. The number varies
B. ---------
C. ---------
D. ---------
~~

G2E03 (C)
What part of a data packet contains the routing and handling information?
A. ---------
B. ---------
C. Header
D. ---------
~~

# SIMPLIFIED STUDY GUIDE FOR GENRAL CLASS...

G2E04 (B)
Which of the following 20 meter band segments is most often used for most data transmissions?
A. ---------
B. 14.070 - 14.100 MHz
C. ---------
D. ---------
~~

G2E05 (C)
Which of the following describes Baudot RTTY?
A. ---------
B. ---------
C. 5-bit code, with additional start and stop bits
D. ---------
~~

G2E06 (B)
What is the most common frequency shift for RTTY emissions in the amateur HF
bands?
A. ---------
B. 170 Hz
C. ---------
D. ---------
~~

G2E07 (B)
What does the abbreviation "RTTY" stand for?
A. ---------
B. Radio-Teletype
C. ---------
D. ---------
~~

G2E08 (A)
What segment of the 80 meter band is most commonly used for data transmissions?
A. 3570 – 3600 kHz
B. ---------
C. ---------
D. ---------
~~

G2E09 (D)
Where are PSK signals generally found on the 20 meter band?
A. ---------
B. ---------
C. ---------
D. Around 14.070 MHz
~~

G2E10 (D)
What is a major advantage of MFSK16 compared to other digital modes?
A. ---------
B. ---------
C. ---------
D. It offers good performance in weak signal environment without error correction
~~

G2E11 (B)
What does the abbreviation "MFSK" stand for?
A. ---------
B. Multi (or Multiple) Frequency Shift Keying
C. ---------
D. ---------
~~


G2F - CW operating procedures and procedural signals, Q signals and common abbreviations; full break in


G2F01 (D)
Which of the following describes full break-in telegraphy (QSK)?
A. ---------
B. ---------
C. ---------
D. Incoming signals are received between transmitted code character elements
~~

# SIMPLIFIED STUDY GUIDE FOR GENRAL CLASS…

G2F02 (A)
What should you do if a CW station sends "QRS" when using Morse code?
A. Send slower
B. ---------
C. ---------
D. ---------
~~

G2F03 (C)
What does it mean when a CW operator sends "KN" at the end of a transmission?
A. ---------
B. ---------
C. Listening only for a specific station or stations
D. ---------
~~

G2F04 (D)
What does it mean when a CW operator sends "CL" at the end of a transmission?
A. ---------
B. ---------
C. ---------
D. Closing station
~~

G2F05 (B)
What is the best speed to use answering a CQ in Morse Code?
A. ---------
B. The speed at which the CQ was sent
C. ---------
D. ---------
~~

G2F06 (D)
What does the term "zero beat" mean in CW operation?
A. ---------
B. ---------
C. ---------
D. Matching the frequency of the transmitting station
~~

G2FO7 (A)
When sending CW, what does a "C" mean when added to the RST report?
A. Chirpy or unstable signal
B. ---------
C. ---------
D. ---------
~~

G2F08 (C)
What prosign is sent using CW to indicate the end of a formal message?
A. ---------
B. ---------
C. AR
D. ---------
~~

G2F09 (C)
What does the Q signal "QSL" mean when operating CW?
A. ---------
B. ---------
C. I acknowledge receipt
D. ---------
~~

G2F10 (B)
What does the Q signal "QRQ" mean when operating CW?
A. ---------
B. Send faster
C. ---------
D. ---------
~~

G2F11 (D)
What does the Q signal "QRV" mean when operating CW?
A. ---------
B. ---------
C. ---------
D. I am ready to receive messages
~~

# PROPAGATION

SUBELEMENT G3 - RADIO WAVE PROPAGATION [3 Exam Questions - 3 Groups]

G3A - Sunspots and solar radiation; ionospheric disturbances; propagation forecasting and indices

G3A01 (A)
What can be done at an amateur station to continue communications during a sudden ionospheric disturbance?
A. Try a higher frequency
B. ---------
C. ---------
D. ---------
~~
G3A02 (B)
What effect does a Sudden Ionospheric Disturbance (SID) have on the daytime ionospheric propagation of HF radio waves?
A. ---------
B. It disrupts signals on lower frequencies more than those on higher frequencies
C. ---------
D. ---------

# SIMPLIFIED STUDY GUIDE FOR GENRAL CLASS...

G3A03 (C)
How long does it take the increased ultraviolet and X-ray radiation from solar flares to affect radio-wave propagation on the Earth?
A. ---------
B. ---------
C. Approximately 8 minutes
D. ---------
~~

G3A04 (B)
What is measured by the solar flux index?
A. ---------
B. The radio energy emitted by the sun
C. ---------
D. ---------
~~

G3A05 (D)
What is the solar-flux index?
A. ---------
B. ---------
C. ---------
D. A measure of solar activity at 10.7 cm

~~

G3A06 (D)
What is a geomagnetic disturbance?
A. ---------
B. ---------
C. ---------
D. A significant change in the Earth's magnetic field over a short period of time
~~

G3A07 (A)
Which latitudes have propagation paths that are more sensitive to geomagnetic
disturbances?
A. Those greater than 45 degrees North or South latitude
B. ---------
C. ---------
D. ---------

G3A08 (B)
What can be an effect of a geomagnetic storm on radio-wave propagation?
A. ---------
B. Degraded high-latitude HF propagation
C. ---------
D. ---------

G3A09 (C)
What is the effect on radio communications when sunspot numbers are high?
A. ---------
B. ---------
C. Long-distance communication in the upper HF and lower VHF range is enhanced
D. ---------

G3A10 (A)
What is the sunspot number?
A. A measure of solar activity based on counting sunspots and sunspot groups
B. ---------
C. ---------
D. ---------

G3A11 (D)
How long is the typical sunspot cycle?
A. ---------
B. ---------
C. ---------
D. Approximately 11 years

G3A12 (B)
What is the K-index?
A. ---------
B. A measure of the short term stability of the Earth's magnetic field
C. ---------
D. ---------

# SIMPLIFIED STUDY GUIDE FOR GENRAL CLASS...

~~

G3A13 (C)
What is the A-index?
A. ---------
B. ---------
C. An indicator of the long term stability of the Earth's geomagnetic field
D. ---------
~~

G3A14 (B)
How are radio communications usually affected by the charged particles that reach the Earth from solar coronal holes?
A. ---------
B. HF communications are disturbed
C. ---------
D. ---------
~~

G3A15 (D)
How long does it take charged particles from Coronal Mass Ejections to affect radio-wave propagation on the Earth?
A. ---------
B. ---------
C. ---------
D. 20 to 40 hours
~~

G3A16 (A)
What is a possible benefit to radio communications resulting from periods of high geomagnetic activity?
A. Aurora that can reflect VHF signals
B. ---------
C. ---------
D. ---------
~~

G3A17 (D)
At what point in the solar cycle does the 20 meter band usually support worldwide propagation during daylight hours?
A. ---------
B. ---------
C. ---------
D. At any point in the solar cycle

G3A18 (C)
If the HF radio-wave propagation (skip) is generally good on the 24-MHz and 28-MHz bands for several days, when might you expect a similar condition to occur?
A. ---------
B. ---------
C. 28 days later
D. ---------

G3A19 (D)
Which frequencies are least reliable for long distance communications during periods of low solar activity?
A. ---------
B. ---------
C. ---------
D. Frequencies above 20 MHz

G3B - Maximum Usable Frequency; Lowest Usable Frequency; propagation "hops"

G3B01 (B)

Which band should offer the best chance for a successful contact if the maximum usable frequency (MUF) between the two stations is 22 MHz?
A. ---------
B. 15 meters
C. ---------
D. ---------

G3B02 (C)
Which band should offer the best chance for a successful contact if the maximum usable frequency (MUF) between the two stations is 16 MHz?
A. ---------
B. ---------
C. 20 meters
D. ---------
~~

G3B03 (A)
Which of the following guidelines should be selected for lowest attenuation when transmitting on HF?
A. Select a frequency just below the MUF
B. ---------
C. ---------
D. ---------
~~

G3B04 (A)
What is a reliable way to determine if the maximum usable frequency (MUF) is high enough to support 28-MHz propagation between your station and Western Europe?
A. Listen for signals on a 28 MHz international beacon
B. ---------
C. ---------
D. ---------
~~

G3B05 (A)
What usually happens to radio waves with frequencies below the maximum usable frequency (MUF) when they are sent into the ionosphere?
A. They are bent back to the Earth
B. ---------
C. ---------
D. ---------
~~

G3B06 (C)
What usually happens to radio waves with frequencies below the lowest usable frequency (LUF)?
A. ---------
B. ---------
C. They are completely absorbed by the ionosphere
D. ---------
~~

G3B07 (A)
What does LUF stand for?
A. The Lowest Usable Frequency for communications between two points
B. ---------
C. ---------
D. ---------
~~

G3B08 (B)
What does MUF stand for?
A. ---------
B. The Maximum Usable Frequency for communications between two points
C. ---------
D. ---------
~~

G3B09 (C)
What is the maximum distance along the Earth's surface that is normally covered in one hop using the F2 region?
A. ---------
B. ---------
C. 2,500 miles
D. ---------
~~
G3B10 (B)
What is the maximum distance along the Earth's surface that is normally covered in one hop using the E region?
A. ---------
B. 1,200 miles
C. ---------
D. ---------
~~

# SIMPLIFIED STUDY GUIDE FOR GENRAL CLASS...

G3B11 (A)
What happens to HF propagation when the lowest usable frequency (LUF) exceeds the maximum usable frequency (MUF)?
A. No HF radio frequency will support communications over the path
B. ---------
C. ---------
D. ---------
~~

G3B12 (D)
What factors affect the maximum usable frequency (MUF)?
A. ---------
B. ---------
C. ---------
D. All of these choices are correct
~~

G3B13 (D)
How might a sky-wave signal sound if it arrives at your receiver by both short path and long path propagation?
A. ---------
B. ---------
C. ---------
D. A well-defined echo can be heard
~~

G3B14 (A)
Which of the following is a good indicator of the possibility of sky-wave propagation on the 6 meter band?
A. Short hop sky-wave propagation on the 10 meter band
B. ---------
C. ---------
D. ---------
~~

G3C - Ionospheric layers; critical angle and frequency; HF scatter; Near Vertical Incidence Sky waves

G3C01 (A)
Which of the following ionospheric layers is closest to the surface of the Earth?
A. The D layer
B. ---------
C. ---------
D. ---------
~~

G3C02 (A)
When can the F2 region be expected to reach its maximum height at your location?
A. At noon during the summer
B. ---------
C. ---------
D. ---------
~~

G3C03 (C)
Why is the F2 region mainly responsible for the longest distance radio wave propagation?
A. ---------
B. ---------
C. Because it is the highest ionospheric region
D. ---------
~~

G3C04 (D)
What does the term "critical angle" mean as used in radio wave propagation?
A. ---------
B. ---------
C. ---------
D. ---------
~~

# SIMPLIFIED STUDY GUIDE FOR GENRAL CLASS...

G3C05 (C)
Why is long distance communication on the 40, 60, 80 and 160 meter bands more difficult during the day?
A. ---------
B. ---------
C. The D layer absorbs these frequencies during daylight hours
D. ---------
~~

G3C06 (B)
What is a characteristic of HF scatter signals?
A. ---------
B. They have a wavering sound
C. ---------
D. ---------
~~

G3C07 (D)
What makes HF scatter signals often sound distorted?
A. ---------
B. ---------
C. ---------
D. Energy is scattered into the skip zone through several radio wave paths
~~

G3C08 (A)
Why are HF scatter signals in the skip zone usually weak?
A. Only a small part of the signal energy is scattered into the skip zone
B. ---------
C. ---------
D. ---------
~~

G3C09 (B)
What type of radio wave propagation allows a signal to be detected at a distance too far for ground wave propagation but too near for normal sky wave propagation?
A. ---------
B. Scatter
C. ---------
D. ---------
~~

G3C10 (D)
Which of the following might be an indication that signals heard on the HF bands are being received via scatter propagation?
A. ---------
B. ---------
C. ---------
D. The signal is heard on a frequency above the maximum usable frequency
~~

G3C11 (A)
Which of the following is true about ionospheric absorption near the maximum usable frequency (MUF)?
A. Absorption will be minimum
B. ---------
C. ---------
D. ---------
~~

G3C12 (D)
Which ionospheric layer is the most absorbent of long skip signals during daylight hours on frequencies below 10 MHz?
A. ---------
B. ---------
C. ---------
D. The D layer
~~

G3C13 (B)
What is Near Vertical Incidence Sky-wave (NVIS) propagation?
A. ---------
B. Short distance HF propagation using high elevation angles
C. ---------
D. ---------

# SIMPLIFIED STUDY GUIDE FOR GENRAL CLASS...

~~

G3C14 (B)
Which of the following antennas will be most effective for skip communications on 40 meters during the day?
A. ---------
B. A horizontal dipole placed between 1/8 and 1/4 wavelength above the ground
C. ---------
D. ---------
~~

# AMATEUR PRACTICES

G4 - AMATEUR RADIO PRACTICES [5 Questions - 5 groups]

G4A - Two-tone Test; amplifier tuning and neutralization; DSP

G4A01 (B)
Which of the following is one use for a DSP in an amateur station?
A. ---------
B. To remove noise from received signals
C. ---------
D. ---------
~~

G4A02 (B)
Which of the following instruments may be used to measure the output of a single-sideband transmitter when performing a two-tone test of amplitude linearity?
A. ---------
B. An oscilloscope
C. ---------

D. ---------
~~

G4A03 (D)
Which of the following is needed for a DSP IF filter?
A. An Analog to Digital Converter
B. Digital to Analog Converter
C. A Digital Processor Chip
D. All of the these answers are correct
~~

G4A04 (A)
Which of the following is an advantage of a receiver IF filter created with a DSP as compared to an analog filter?
A. A wide range of filter bandwidths and shapes can be created
B. ---------
C. ---------
D. ---------
~~

G4A05 (B)
How is DSP filtering accomplished?
A. ---------
B. By converting the signal from analog to digital and using digital processing
C. ---------
D. ---------
~~

G4A06 (B)
What reading on the plate current meter of a vacuum tube RF power amplifier indicates correct adjustment of the plate tuning control?
A. ---------
B. A pronounced dip
C. ---------
D. ---------
~~

G4A07 (D)
What is the correct adjustment for the "Load" or "Coupling" control of a vacuum tube RF power amplifier?
A. ---------
B. ---------
C. ---------
D. Maximum power output without exceeding maximum allowable plate current
~~

G4A08 (C)
Which of the following techniques is used to neutralize an RF amplifier?
A. ---------
B. ---------
C. Negative feedback
D. ---------
~~

G4A09 (B)
What does a neutralizing circuit do in an RF amplifier?
A. ---------
B. It cancels the effects of positive feedback
C. ---------
D. ---------
~~

G4A10 (B)
What is the reason for neutralizing the final amplifier stage of a transmitter?
A. ---------
B. To eliminate self oscillations
C. ---------
D. ---------
~~

G4A11 (A)
What type of transmitter performance does a two-tone test analyze?
A. Linearity
B. ---------
C. ---------
D. ---------
~~

G4A12 (B)
What type of signals are used to conduct a two-tone test?
A. ---------
B. Two non-harmonically related audio signals
C. ---------
D. ---------
~~

G4A13 (B)
Which of the following performs automatic notching of interfering carriers?
A. ---------
B. A DSP filter
C. ---------
D. ---------
~~

G4B - Test and monitoring equipment

G4B01 (D)
What item of test equipment contains horizontal and vertical channel amplifiers?
A. ---------
B. ---------
C. ---------
D. An oscilloscope
~~

G4B02 (D)
Which of the following is an advantage of an oscilloscope versus a digital voltmeter?
A. ---------
B. ---------
C. ---------
D. Complex waveforms can be measured
~~

G4B03 (D)
How would a signal tracer normally be used?
A. ---------
B. ---------
C. ---------
D. To identify an inoperative stage in a receiver
~~

G4B04 (C)
How is a noise bridge normally used?
A. ---------
B. ---------
C. It is connected between a receiver and an antenna of unknown impedance and is
   adjusted for minimum noise
D. ---------
~~

G4B05 (A)
Which of the following is the best instrument to use to check the keying waveform of a CW transmitter?
A. A monitoring oscilloscope
B. ---------
C. ---------
D. ---------
~~

G4B06 (D)
What signal source is connected to the vertical input of a monitoring oscilloscope when checking the quality of a transmitted signal?
A. ---------
B. ---------
C. ---------
D. The attenuated RF output of the transmitter
~~

G4B07 (C)
What is an advantage of a digital voltmeter as compared to an analog voltmeter?
A. ---------
B. ---------
C. Significantly better precision for most uses
D. ---------
~~

# SIMPLIFIED STUDY GUIDE FOR GENRAL CLASS...

G4B08 (A)
What instrument may be used to monitor relative RF output when making
antenna and transmitter adjustments?
A. A field-strength meter
B. ---------
C. ---------
D. ---------
~~

G4B09 (C)
How much must the power output of a transmitter be raised to change the "S" meter reading on a distant receiver from S8 to S9?
A. ---------
B. ---------
C. Approximately 4 times
D. ---------
~~

G4B10 (B)
Which of the following can be determined with a field strength meter?
A. ---------
B. The radiation pattern of an antenna
C. ---------
D. ---------
~~

G4B11 (A)
Which of the following might be a use for a field strength meter?
A. Close-in radio direction-finding
B. ---------
C. ---------
D. ---------
~~

G4B12 (B)
What is one way a noise bridge might be used?
A. ---------
B. Pre-tuning an antenna tuner
C. ---------
D. ---------
~~

G4B13 (A)
What is one measurement that can be made with a dip meter?
A. The resonant frequency of a circuit
B. ---------
C. ---------
D. ---------
~~

G4B14 (C)
Which of the following must be connected to an antenna analyzer when it is being used for SWR measurements?
A. ---------
B. ---------
C. Antenna and feedline
D. ---------
~~

G4B15 (A)
Which of the following can be measured with a directional wattmeter?
A. Standing Wave Ratio
B. ---------
C. ---------
D. ---------
~~

G4B16 (D)
Why is high input impedance desirable for a voltmeter?
A. ---------
B. ---------
C. ---------
D. It decreases the loading on circuits being measured
~~


G4C - Interference with consumer electronics; grounding

G4C01 (B)
Which of the following might be useful in reducing RF interference to audio-frequency devices?
A. ---------
B. Bypass capacitor
C. ---------
D. ---------

# SIMPLIFIED STUDY GUIDE FOR GENRAL CLASS...

~~

G4C02 (B)
Which of the following should be installed if a properly operating amateur station is interfering with a nearby telephone?
A. ---------
B. An RFI filter at the affected telephone
C. ---------
D. ---------
~~

G4C03 (C)
What sound is heard from a public-address system if there is interference from a nearby single-sideband phone transmitter?
A. ---------
B. ---------
C. Distorted speech
D. ---------
~~

G4C04 (A)
What is the effect on a public-address system if there is interference from nearby CW transmitter?
A. On-and-off humming or clicking
B. ---------
C. ---------
D. ---------
~~

G4C05 (D)
What might be the problem if you receive an RF burn when touching your equipment while transmitting on a HF band, assuming the equipment is connected to a ground rod?
A. ---------
B. ---------
C. ---------
D. The ground wire is resonant
~~

G4C06 (D)
Which of the following is an important reason to have a good station ground?
A. To reduce the likelihood of RF burns
B. To reduce the likelihood of electrical shock
C. To reduce interference
D. All of these answers are correct
~~

G4C07 (A)
What is one good way to avoid stray RF energy in an amateur station?
A. Keep the station's ground wire as short as possible
B. ---------
C. ---------
D. ---------
~~

G4C08 (A)
Which of the following is a reason to place ferrite beads around audio cables to reduce common mode RF interference?
A. They act as a series inductor
B. ---------
C. ---------
D. ---------
~~

G4C09 (C)
Which of the following statements about station grounding is true?
A. ---------
B. ---------
C. RF hot spots can occur in a station located above the ground floor if the
   equipment is grounded by a long ground wire
D. ---------
~~

G4C10 (C)
Which of the following is covered in the National Electrical Code?
A. ---------
B. ---------
C. Electrical safety inside the ham shack
D. ---------
~~

G4C11 (A)
Which of the following can cause unintended rectification of RF signal energy and can result in interference to your station as well as nearby radio and TV receivers?
A. Induced currents in conductors that are in poor electrical contact
B. ---------
C. ---------
D. ---------
~~

G4C12 (C)
What is one cause of broadband radio frequency interference at an amateur radio station?
A. ---------
B. ---------
C. Arcing at a poor electrical connection
D. ---------
~~

G4C13 (D)
How can a ground loop be avoided?
A. ---------
B. ---------
C. ---------
D. Connect all ground conductors to a single point
~~


**G4D** - Speech processors; S meters; common connectors


G4D01 (D)
What is the reason for using a properly adjusted speech processor with a single sideband phone transmitter?
A. ---------
B. ---------
C. ---------
D. It improves signal intelligibility at the receiver
~~

G4D02 (B)
Which of the following describes how a speech processor affects a transmitted single sideband signal?
A. ---------
B. It increases the average power
C. ---------
D. ---------
~~

G4D03 (D)
Which of the following can be the result of an incorrectly adjusted speech processor?
A. Distorted speech
B. Splatter
C. Excessive background pickup
D. All of these answers are correct
~~

G4D04 (C)
What does an S-meter measure?
A. ---------
B. ---------
C. Received signal strength
D. ---------
~~

G4D05 (D)
How does an S-meter reading of 20 db over S-9 compare to an S-9 signal, assuming a properly calibrated S meter?
A. ---------
B. ---------
C. ---------
D. It is 100 times stronger
~~

G4D06 (A)
Where is an S-meter generally found?
A. In a receiver
B. ---------
C. ---------
D. ---------
~~

# SIMPLIFIED STUDY GUIDE FOR GENRAL CLASS...

G4D07 (A)
Which of the following describes a Type-N connector?
A. A moisture resistant RF connector useful to 10 GHz
B. ---------
C. ---------
D. ---------
~~

G4D08 (D)
Which of the following connectors would be a good choice for a serial data port?
A. ---------
B. ---------
C. ---------
D. DB-9
~~

G4D09 (C)
Which of these connector types is commonly used for RF service at frequencies up to 150 MHz?
A. ---------
B. ---------
C. UHF
D. ---------
~~

G4D10 (C)
Which of these connector types is commonly used for audio signals in amateur radio stations?
A. ---------
B. ---------
C. RCA Phono
D. ---------
~~

G4D11 (B)
What is the main reason to use keyed connectors over non-keyed types?
A. ---------
B. Reduced chance of damage due to incorrect mating
C. ---------
D. ---------
~~

G4E - HF mobile radio installations; emergency and battery powered operation

G4E01 (D)
Which of the following emission types are permissible while operating HF mobile?
A. CW
B. SSB
C. FM
D. All of these choices are correct
~~

G4E02 (C)
What is alternator whine?
A. ---------
B---------
C. A tone or buzz in transmitted or received audio that varies with engine speed
D. ---------
~~

G4E03 (A)
Which of the following power connections would be the best for a 100-watt HF mobile installation?
A. A direct, fused connection to the battery using heavy gauge wire
B. ---------
C. ---------
D. ---------
~~

G4E04 (B)
Why is it best NOT to draw the DC power for a 100-watt HF transceiver from an automobile's cigarette lighter socket?
A. ---------
B. The socket's wiring may be inadequate for the current being drawn by the
   transceiver
C. ---------
D. ---------
~~

G4E05 (C)
Which of the following most limits the effectiveness of an HF mobile transceiver operating in the 75 meter band?
A. ---------
B. ---------
C. The HF mobile antenna system
D. ---------
~~

G4E06 (A)
Which of the following is true of an emergency generator installation?
A. The generator should be located in a well ventilated area
B. ---------
C. ---------
D. ---------
~~

G4E07 (C)
When might a lead-acid storage battery give off explosive hydrogen gas?
A. ---------
B. ---------
C. When being charged
D. ---------
~~

G4E08 (A)
What is the name of the process by which sunlight is changed directly into electricity?
A. Photovoltaic conversion
B. ---------
C. ---------
D. ---------
~~

G4E09 (B)
What is the approximate open-circuit voltage from a modern, well illuminated
photovoltaic cell?
A. ---------
B. 0.5 VDC
C. ---------
D. ---------
~~

G4E10 (A)
Which of these materials is used as the active element of a solar cell?
A. Doped Silicon
B. ---------
C. ---------
D. ---------
~~

G4E11 (C)
Which of the following is a disadvantage to using wind power as the primary source of power for an emergency station?
A. ---------
B. ---------
C. A large energy storage system is needed to supply power when the wind is not
   blowing
D. ---------
~~

G4E12 (A)
Which of the following is a primary reason for not placing a gasoline-fueled generator inside an occupied area?
A. Danger of carbon monoxide poisoning
B. ---------
C. ---------
D. ---------
~~

G4E13 (A)
Why would it be unwise to power your station by back feeding the output of a gasoline generator into your house wiring by connecting the generator through an AC wall outlet?
A. It might present a hazard for electric company workers
B. ---------
C. ---------
D. ---------
~~

# ELECTRICAL PRINCIPLES

SUBELEMENT G5 – ELECTRICAL PRINCIPLES [3 exam questions – 3 groups]

G5A - Resistance; reactance; inductance; capacitance; impedance; impedance matching

G5A01 (C)
What is impedance?
A. ---------
B. ---------
C. The opposition to the flow of current in an AC circuit
D. ---------
~~

G5A02 (B)
What is reactance?
A. ---------
B. Opposition to the flow of alternating current caused by capacitance or inductance
C. ---------
D. ---------
~~

# SIMPLIFIED STUDY GUIDE FOR GENRAL CLASS...

G5A03 (D)
Which of the following causes opposition to the flow of alternating current in an inductor?
A. ---------
B. ---------
C. ---------
D. Reactance
~~

G5A04 (C)
Which of the following causes opposition to the flow of alternating current in a capacitor?
A. ---------
B. ---------
C. Reactance
D. ---------
~~

G5A05 (D)
How does a coil react to AC?
A. ---------
B. ---------
C. ---------
D. As the frequency of the applied AC increases, the reactance increases
~~

G5A06 (A)
How does a capacitor react to AC?
A. As the frequency of the applied AC increases, the reactance decreases
B. ---------
C. ---------
D. ---------
~~

G5A07 (D)
What happens when the impedance of an electrical load is equal to the internal
impedance of the power source?
A. ---------
B. ---------
C. ---------
D. The source can deliver maximum power to the load

G5A08 (A)
Why is impedance matching important?
A. So the source can deliver maximum power to the load
B. ---------
C. ---------
D. ---------

G5A09 (B)
What unit is used to measure reactance?
A. ---------
B. Ohm
C. ---------
D. ---------

G5A10 (B)
What unit is used to measure impedance?
A. ---------
B. Ohm
C. ---------
D. ---------

G5A11 (A)
Why should core saturation of a conventional impedance matching transformer be avoided?
A. Harmonics and distortion could result
B. ---------
C. ---------
D. ---------

G5A12 (B)
What is one reason to use an impedance matching transformer?
A. ---------
B. To maximize the transfer of power
C. ---------
D. ---------

# SIMPLIFIED STUDY GUIDE FOR GENRAL CLASS...

G5A13 (D)
Which of the following devices can be used for impedance matching at radio frequencies?
A. ---------
B. ---------
C. ---------
D. All of these choices are correct
~~

G5A14 (A)
Which of the following describes one method of impedance matching between two AC circuits?
A. Insert an LC network between the two circuits
B. ---------
C. ---------
D. ---------
~~

G5B - The Decibel; current and voltage dividers; electrical power calculations; sine wave root-mean-square (RMS) values; PEP calculations

G5B01 (B)
A two-times increase or decrease in power results in a change of how many dB?
A. ---------
B. 3 dB
C. ---------
D. ---------
~~

G5B02 (C)
How does the total current relate to the individual currents in each branch of a parallel circuit?
A. ---------
B. ---------
C. It equals the sum of the currents through each branch
D. ---------
~~

G5B03 (B)
How many watts of electrical power are used if 400 VDC is supplied to an 800-ohm load?
A. ---------
B. 200 watts
C. ---------
D. ---------
~~

G5B04 (A)
How many watts of electrical power are used by a 12-VDC light bulb that draws 0.2 amperes?
A. 2.4 watts
B. ---------
C. ---------
D. ---------
~~

G5B05 (A)
How many watts are being dissipated when a current of 7.0 milliamperes flows through 1.25 kilohms?
A. Approximately 61 milliwatts
B. ---------
C. ---------
D. ---------
~~

G5B06 (B)
What is the output PEP from a transmitter if an oscilloscope measures 200 volts peak-to-peak across a 50-ohm dummy load connected to the transmitter output?
A. ---------
B. 100 watts
C. ---------
D. ---------
~~

G5B07 (C)
Which measurement of an AC signal is equivalent to a DC voltage of the same value?
A. ---------
B. ---------
C. The RMS value
D. ---------

# SIMPLIFIED STUDY GUIDE FOR GENRAL CLASS...

~~

G5B08 (D)
What is the peak-to-peak voltage of a sine wave that has an RMS voltage of 120 volts?
A. ---------
B. ---------
C. ---------
D. 339.4 volts
~~

G5B09 (B)
What is the RMS voltage of sine wave with a value of 17 volts peak?
A. ---------
B. 12 volts
C. ---------
D. ---------
~~

G5B11 (B)
What is the ratio of peak envelope power to average power for an un-modulated carrier?
A. ---------
B. 1.00
C. ---------
D. ---------
~~

G5B12 (B)
What would be the voltage across a 50-ohm dummy load dissipating 1200 watts?
A. ---------
B. 245 volts
C. ---------
D. ---------
~~

G5B13 (C)
What percentage of power loss would result from a transmission line loss of 1 dB?
A. ---------
B. ---------
C. 20.5 %
D. ---------
~~

G5B14 (B)
What is the output PEP from a transmitter if an oscilloscope measures 500 volts peak-to-peak across a 50-ohm resistor connected to the transmitter output?
A. ---------
B. 625 watts
C. ---------
D. ---------
~~

G5B15 (B)
What is the output PEP of an unmodulated carrier if an average reading wattmeter connected to the transmitter output indicates 1060 watts?
A. ---------
B. 1060 watts
C. ---------
D. ---------
~~


G5C – Resistors, capacitors, and inductors in series and parallel; transformers


G5C01 (C)
What causes a voltage to appear across the secondary winding of a transformer when an AC voltage source is connected across its primary winding?
A. ---------
B. ---------
C. Mutual inductance
D. ---------
~~

G5C02 (B)
Where is the source of energy normally connected in a transformer?
A. ---------
B. To the primary winding
C. ---------
D. ---------
~~

G5C03 (A)
What is current in the primary winding of a transformer called if no load is attached to the secondary?
A. Magnetizing current
B. ---------
C. ---------
D. ---------
~~

G5C04 (C)
What is the total resistance of three 100-ohm resistors in parallel?
A. ---------
B. ---------
C. 33.3 ohms
D. ---------
~~

G5C05 (C)
What is the value of each resistor if three equal value resistors in parallel produce 50 ohms of resistance, and the same three resistors in series produce 450 ohms?
A. ---------
B. ---------
C. 150 ohms
D. ---------
~~

G5C06 (C)
What is the voltage across a 500-turn secondary winding in a transformer if the 2250-turn primary is connected to 120 VAC?
A. ---------
B. ---------
C. 26.7 volts
D. ---------
~~

G5C07 (A)
What is the turns ratio of a transformer used to match an audio amplifier having a 600-ohm output impedance to a speaker having a 4-ohm impedance?
A. 12.2 to 1
B. ---------
C. ---------
D. ---------
~~

G5C08 (D)
What is the equivalent capacitance of two 5000 picofarad capacitors and one 750 picofarad capacitor connected in parallel?
A. ---------
B. ---------
C. ---------
D. 10750 picofarads
~~

G5C09 (C)
What is the capacitance of three 100 microfarad capacitors connected in series?
A. ---------
B. ---------
C. 33.3 microfarads
D. ---------
~~

G5C10 (C)
What is the inductance of three 10 millihenry inductors connected in parallel?
A. ---------
B. ---------
C. 3.3 millihenrys
D. ---------
~~

G5C11 (C)
What is the inductance of a 20 millihenry inductor in series with a 50 millihenry inductor?
A. ---------
B. ---------
C. 70 millihenrys
D. ---------

# SIMPLIFIED STUDY GUIDE FOR GENRAL CLASS...

~~
G5C12 (B)
What is the capacitance of a 20 microfarad capacitor in series with a 50 microfarad capacitor?
A. ---------
B. 14.3 microfarads
C. ---------
D. ---------
~~

G5C13 (C)
What component should be added to a capacitor in a circuit to increase the circuit capacitance?
A. ---------
B. ---------
C. A capacitor in parallel
D. ---------
~~

G5C14 (D)
What component should be added to an inductor in a circuit to increase the circuit inductance?
A. ---------
B. ---------
C. ---------
D. An inductor in series
~~

G5C15 (A)
What is the total resistance of a 10 ohm, a 20 ohm, and a 50 ohm resistor in parallel?
A. 5.9 ohms
B. ---------
C. ---------
D. ---------
~~

G5C16 (B)
What component should be added to an existing resistor in a circuit to increase circuit resistance?
A. ---------
B. A resistor in series
C. ---------
D. ---------

# CIRCUIT COMPONENTS

SUBELEMENT G6 – CIRCUIT COMPONENTS [3 exam question – 3 groups]

G6A - Resistors; capacitors; inductors

G6A01 (C)
What will happen to the resistance if the temperature of a carbon resistor is increased?
A. ---------
B. ---------
C. It will change depending on the resistor's temperature coefficient rating
D. ---------
~~

G6A02 (D)
What type of capacitor is often used in power-supply circuits to filter the rectified AC?
A. ---------
B. ---------
C. ---------
D. Electrolytic
~~

G6A03 (D)
Which of the following is the primary advantage of ceramic capacitors?
A. ---------
B. ---------
C. ---------
D. Comparatively low cost

~~

G6A04 (C)
Which of the following is an advantage of an electrolytic capacitor?
A. ---------
B. ---------
C. High capacitance for given volume
D. ---------

~~

G6A05 (A)
Which of the following is one effect of lead inductance in a capacitor used at VHF and above?
A. Effective capacitance may be reduced
B. ---------
C. ---------
D. ---------

~~

G6A06 (B)
What is the main disadvantage of using a conventional wire-wound resistor in a resonant circuit?
A. ---------
B. The resistor's inductance could detune the circuit
C. ---------
D. ---------

~~

G6A07 (D)
What is an advantage of using a ferrite core with a toroidal inductor?
A. ---------
B. ---------
C. ---------
D. All of these choices are correct

~~

G6A08 (C)
How should two solenoid inductors be placed so as to minimize their mutual inductance?
A. ---------
B. ---------
C. With their winding axes at right angles to each another
D. ---------
~~

G6A09 (B)
Why might it be important to minimize the mutual inductance between two
inductors?
A. ---------
B. To reduce or eliminate unwanted coupling
C. ---------
D. ---------
~~

G6A10 (B)
What is an effect of inter-turn capacitance in an inductor?
A. ---------
B. The inductor may become self resonant at some frequencies
C. ---------
D. ---------
~~

G6A11 (D)
What is the common name for a capacitor connected across a transformer secondary that is used to absorb transient voltage spikes?
A. ---------
B. ---------
C. ---------
D. Suppressor capacitor
~~

G6A12 (D)
What is the common name for an inductor used to help smooth the DC output from the rectifier in a conventional power supply?
A. ---------
B. ---------
C. ---------
D. Filter choke
~~

G6A13 (B)

# SIMPLIFIED STUDY GUIDE FOR GENRAL CLASS...

What type of component is a thermistor?
A. ---------
B. A device having a controlled change in resistance with temperature variations
C. ---------
D. ---------
~~

G6B - Rectifiers; solid state diodes and transistors; solar cells; vacuum tubes; batteries

G6B01 (C)
What is the peak-inverse-voltage rating of a rectifier?
A. ---------
B. ---------
C. The maximum voltage the rectifier will handle in the non-conducting direction
D. ---------
~~

G6B02 (A)
What are the two major ratings that must not be exceeded for silicon-diode rectifiers?
A. Peak inverse voltage; average forward current
B. ---------
C. ---------
D. ---------
~~

G6B03 (B)
What is the approximate junction threshold voltage of a germanium diode?
A. ---------
B. 0.3 volts
C. ---------
D. ---------
~~

G6B04 (C)

When two or more diodes are connected in parallel to increase current handling capacity, what is the purpose of the resistor connected in series with each diode?
A. ---------
B. ---------
C. The resistors ensure that one diode doesn't carry most of the current
D. ---------
~~

G6B05 (C)
What is the approximate junction threshold voltage of a silicon diode?
A. ---------
B. ---------
C. 0.7 volts
D. ---------
~~

G6B06 (A)
Which of the following is an advantage of using a Schottky diode in an RF switching circuit as compared to a standard silicon diode?
A. Lower capacitance
B. ---------
C. ---------
D. ---------
~~

G6B07 (A)
What are the stable operating points for a bipolar transistor that is used as a switch in a logic circuit?
A. Its saturation and cut-off regions
B. ---------
C. ---------
D. ---------
~~

G6B08 (D)
Why is it often necessary to insulate the case of a large power transistor?
A. ---------
B. ---------
C. ---------
D. To avoid shorting the collector or drain voltage to ground
~~

G6B09 (B)
Which of the following describes the construction of a MOSFET?
A. ---------
B. The gate is separated from the channel with a thin insulating layer
C. ---------
D. ---------
~~

G6B10 (A)
Which element of a triode vacuum tube is used to regulate the flow of electrons between cathode and plate?
A. Control grid
B. ---------
C. ---------
D. ---------
~~

G6B11 (B)
Which of the following solid state devices is most like a vacuum tube in its general characteristics?
A. A bipolar transistor
B. ---------
C. ---------
D. ---------
~~

G6B12 (A)
What is the primary purpose of a screen grid in a vacuum tube?
A. To reduce grid-to-plate capacitance
B. ---------
C. ---------
D. ---------
~~

G6B13 (B)
What is an advantage of the low internal resistance of Nickel Cadmium batteries?
A. ---------
B. High discharge current
C. ---------
D. ---------
~~

G6B14 (C)
What is the minimum allowable discharge voltage for maximum life of a standard 12 volt lead acid battery?
A. ---------
B. ---------
C. 10.5 volts
D. ---------
~~

G6B15 (D)
When is it acceptable to recharge a carbon-zinc primary cell?
A. ---------
B. ---------
C. ---------
D. Never
~~

G6B16 (C)
Which of the following is a rechargeable battery?
A. ---------
B. ---------
C. Nickel Metal Hydride
D. ---------
~~

G6C - Analog and digital integrated circuits (IC's); microprocessors; memory; I/O devices; microwave IC's (MMIC's ); display devices

G6C01 (D)
Which of the following is most often provided as an analog integrated circuit?
A. ---------
B. ---------
C. ---------
D. Linear voltage regulator
~~

G6C02 (C)
Which of the following is the most commonly used digital logic family of integrated circuits?
A. ---------
B. ---------
C. CMOS
D. ---------

# SIMPLIFIED STUDY GUIDE FOR GENRAL CLASS...

~~
G6C03 (A)
Which of the following is an advantage of CMOS Logic integrated circuits compared to TTL logic circuits?
A. Low power consumption
B. ---------
C. ---------
D. ---------
~~

G6C04 (B)
What is meant by the term ROM?
A. ---------
B. Read Only Memory
C. ---------
D. ---------
~~

G6C05 (C)
What is meant when memory is characterized as "non-volatile"?
A. ---------
B. ---------
C. The stored information is maintained even if power is removed
D. ---------
~~

G6C06 (D)
Which type of integrated circuit is an operational amplifier?
A. ---------
B. ---------
C. ---------
D. ---------
~~

G6C07 (D)
What is one disadvantage of an incandescent indicator compared to a LED?
A. ---------
B. ---------
C. ---------
D. High power consumption
~~

G6C08 (D)
How is an LED biased when emitting light?
A. ---------
B. ---------
C. ---------
D. Forward Biased
~~

G6C09 (A)
Which of the following is a characteristic of a liquid crystal display?
A. It requires ambient or back lighting
B. ---------
C. ---------
D. ---------
~~

G6C10 (B)
What is meant by the term MMIC?
A. ---------
B. Monolithic Microwave Integrated Circuit
C. ---------
D. ---------
~~

G6C11 (B)
What is a microprocessor?
A. ---------
B. A miniature computer on a single integrated circuit chip
C. ---------
D. ---------
~~

G6C12 (A)
What two devices in an amateur radio station might be connected using a USB interface?
A. Computer and transceiver
B. ---------
C. ---------
D. ---------
~~

# 10

# PRACTICAL CIRCUITS

SUBELEMENT G7 – PRACTICAL CIRCUITS [2 exam question – 2 groups]

G7A - Power supplies; transmitters and receivers; filters, schematic drawing symbols

G7A01 (B)
What safety feature does a power-supply bleeder resistor provide?
A. ---------
B. It discharges the filter capacitors
C. ---------
D. ---------
~~

# SIMPLIFIED STUDY GUIDE FOR GENRAL CLASS...

G7A02 (D)
What components are used in a power-supply filter network?
A. ---------
B. ---------
C. ---------
D. Capacitors and inductors
~~

G7A03 (C)
What should be the minimum peak-inverse-voltage rating of the rectifier in a full-wave power supply?
A. ---------
B. ---------
C. Double the normal peak output voltage of the power supply
D. ---------
~~

G7A04 (D)
What should be the approximate minimum peak-inverse-voltage rating of the rectifier in a half-wave power supply?
A. ---------
B. ---------
C. ---------
D. Two times the normal peak output voltage of the power supply
~~

G7A05 (B)
What should be the impedance of a low-pass filter as compared to the impedance of the transmission line into which it is inserted?
A. ---------
B. About the same
C. ---------
D. ---------
~~

G7A06 (B)
Which of the following might be used to process signals from the balanced modulator and send them to the mixer in a single-sideband phone transmitter?
A. ---------
B. Filter
C. ---------
D. ---------
~~

G7A07 (D)
Which circuit is used to combine signals from the carrier oscillator and speech amplifier and send the result to the filter in a typical single-sideband phone transmitter?
A. ---------
B. ---------
C. ---------
D. Balanced modulator
~~

G7A08 (C)
What circuit is used to process signals from the RF amplifier and local oscillator and send the result to the IF filter in a superheterodyne receiver?
A. ---------
B. ---------
C. Mixer
D. ---------
~~

G7A09 (D)
What circuit is used to process signals from the IF amplifier and BFO and send the result to the AF amplifier in a single-sideband phone superheterodyne receiver?
A. ---------
B. ---------
C. ---------
D. Product detector
~~

G7A10 (A)
What is an advantage of a crystal controlled transmitter?
A. Stable output frequency
B. ---------
C. ---------
D. ---------
~~

# SIMPLIFIED STUDY GUIDE FOR GENRAL CLASS...

G7A11 (C)
What is the simplest combination of stages that can be combined to implement a superheterodyne receiver?
A. ---------
B. ---------
C. HF oscillator, mixer, detector
D. ---------
~~

G7A12 (D)
What type of receiver is suitable for CW and SSB reception but does not require a mixer stage or an IF amplifier?
A. ---------
B. ---------
C. ---------
D. A direct conversion receiver

G7A13 (D)
What type of circuit is used in many FM receivers to convert signals coming from the IF amplifier to audio?
A. ---------
B. ---------
C. ---------
D. Discriminator
~~

G7A14 (A)
Which of the following is a desirable characteristic for capacitors used to filter the DC output of a switching power supply?
A. Low equivalent series resistance
B. ---------
C. ---------
D. ---------
~~

G7A15 (C)
Which of the following is an advantage of a switched-mode power supply as compared to a linear power supply?
A. ---------
B. ---------
C. High frequency operation allows the use of smaller components
D. ---------
~~

G7A16(B)
What portion of the AC cycle is converted to DC by a half-wave rectifier?
A. ---------
B. 180 degrees
C. ---------
D. ---------
~~

G7A17 (D)
What portion of the AC cycle is converted to DC by a full-wave rectifier?
A. ---------
B. ---------
C. ---------
D. 360 degrees
~~

G7A18 (A)
What is the output waveform of an unfiltered full-wave rectifier connected to a resistive load?
A. A series of DC pulses at twice the frequency of the AC input
B. ---------
C. ---------
D. ---------
~~

G7A19 (C)
Which symbol in figure G7-1 represents a fixed resistor?
A. ---------
B. ---------
C. Symbol 3
D. ---------
~~

G7A20 (D)
Which symbol in figure G7-1 represents a single cell battery?
A. ---------
B. ---------
C. ---------
D. Symbol 13
~~

# SIMPLIFIED STUDY GUIDE FOR GENRAL CLASS...

G7A21 (B)
Which symbol in figure G7-1 represents a NPN transistor?
A. ---------
B. Symbol 4
C. ---------
D. ---------
~~

G7A22 (C)
Which symbol in figure G7-1 represents a variable capacitor?
A. ---------
B. ---------
C. Symbol 5
D. ---------~

G7A23 (A)
Which symbol in figure G7-1 represents a transformer?
A. Symbol 6
B. ---------
C. ---------
D. ---------
~~

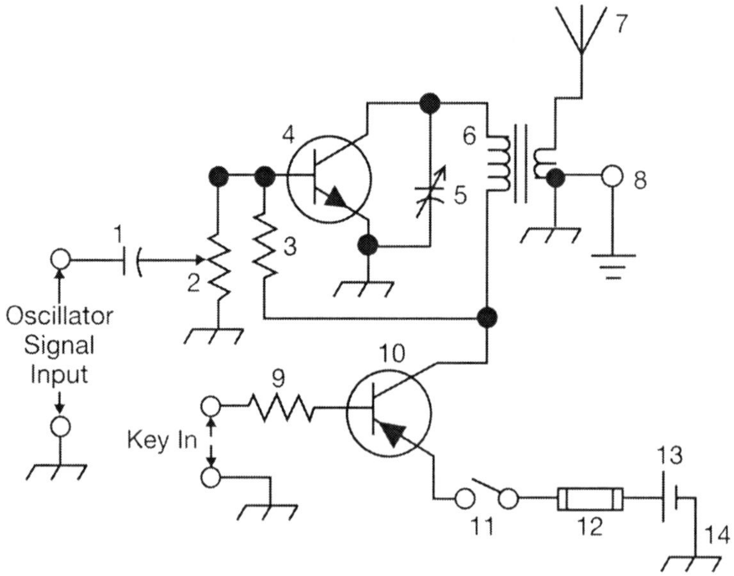

**Figure G7-1 1**

104

G7A24 (C)
Which symbol in figure G7-1 represents a single pole switch?
A. ---------
B. ---------
C. Symbol 11
D. ---------
~~

G7B - Digital circuits (gates, flip-flops, shift registers); amplifiers and oscillators

G7B01 (B)
Which of the following describes a "flip-flop" circuit?
A. ---------
B. A digital circuit with two stable states
C. ---------
D. ---------
~~

G7B02 (A)
Why do digital circuits use the binary number system?
A. Binary "ones" and "zeros" are easy to represent with an "on" or "off" state
B. ---------
C. ---------
D. ---------
~~

G7B03 (C)
What is the output of a two-input NAND gate, given both inputs are "one"?
A. ---------
B. ---------
C. Zero
D. ---------
~~

# SIMPLIFIED STUDY GUIDE FOR GENRAL CLASS...

G7B04 (B)
What is the output of a NOR gate given that both inputs are "zero"?
A. ---------
B. One
C. ---------
D. ---------
~~

G7B05 (C)
How many states are there in a 3-bit binary counter?
A. ---------
B. ---------
C. 8
D. ---------
~~

G7B06 (A)
What is a shift register?
A. A clocked array of circuits that passes data in steps along the array
B. ---------
C. ---------
D. ---------
~~

G7B07 (D)
What are the basic components of virtually all oscillators?
A. ---------
B. ---------
C. ---------
D. A filter and an amplifier operating in a feedback loop
~~

G7B08 (C)
What determines the frequency of an RC oscillator?
A. ---------
B. ---------
C. The phase shift of the RC feedback circuit
D. ---------
~~

G7B09 (C)
What determines the frequency of an LC oscillator?
A. ---------
B. ---------
C. The inductance and capacitance in the tank circuit
D. ---------
~~

G7B10 (D)
Which of the following is a characteristic of a Class A amplifier?
A. ---------
B. ---------
C. ---------
D. Low distortion
~~

G7B11 (B)
For which of the following modes is a Class C power stage appropriate for amplifying a modulated signal?
A. ---------
B. CW
C. ---------
D. ---------
~~

G7B12 (A)
Which of the following is an advantage of a Class C amplifier?
A. High efficiency
B. ---------
C. ---------
D. ---------
~~

G7B13 (B)
How is the efficiency of an RF power amplifier determined?
A. ---------
B. Divide the RF output power by the DC input power
C. ---------
D. ---------
~~

G7B14 (B)
Which of the following describes a linear amplifier?
A. ---------
B. An amplifier whose output preserves the input waveform
C. ---------
D. ---------
~~

# SIGNALS AND EMISSIONS

SUBELEMENT G8 – SIGNALS AND EMISSIONS [2 exam questions – 2 groups]

G8A - Carriers and modulation: AM; FM; single and double sideband ; modulation envelope; deviation; overmodulation

G8A01 (D)
What is the name of the process that changes the envelope of an RF wave to convey information?
A. ---------
B. ---------
C. ---------
D. Amplitude modulation
~~

G8A02 (B)
What is the name of the process that changes the phase angle of an RF wave to convey information?
A. ---------
B. Phase modulation
C. ---------
D. ---------

G8A03 (D)
What is the name of the process which changes the frequency of an RF wave to convey information?
A. ---------
B. ---------
C. ---------
D. Frequency modulation
~~

G8A04 (B)
What emission is produced by a reactance modulator connected to an RF power amplifier?
A. ---------
B. Phase modulation
C. ---------
D. ---------
~~

G8A05 (D)
What type of transmission varies the instantaneous power level of the RF signal to convey information?
A. ---------
B. ---------
C. ---------
D. Amplitude modulation
~~

G8A06 (C)
What is one advantage of carrier suppression in a single-sideband phone
transmission?
A. ---------
B. ---------
C. More transmitter power can be put into the remaining sideband
D. ---------
~~

G8A07 (A)
Which of the following phone emissions uses the narrowest frequency bandwidth?
A. Single sideband
B. ---------
C. ---------
D. ---------

~~

G8A08 (D)
What happens to the signal of an over-modulated single-sideband phone transmitter?
A. ---------
B. ---------
C. ---------
D. It becomes distorted and occupies more bandwidth
~~

G8A09 (B)
What control is typically adjusted for proper ALC setting on an amateur single sideband transceiver?
A. ---------
B. Audio or microphone gain
C. ---------
D. ---------
~~

G8A10 (C)
What is meant by flat-topping of a single-sideband phone transmission?
A. ---------
B. ---------
C. Signal distortion caused by excessive drive
D. ---------
~~

G8A11 (A)
What happens to the RF carrier signal when a modulating audio signal is
applied to an FM transmitter?
A. The carrier frequency changes proportionally to the instantaneous amplitude of the modulating signal
B. ---------
C. --------
D. --------
~~

G8A12 (A)
What signal(s) would be found at the output of a properly adjusted balanced modulator?
A. Both upper and lower sidebands
B. --------
C. --------
D. --------

# SIMPLIFIED STUDY GUIDE FOR GENRAL CLASS...

~~

G8B - Frequency mixing; multiplication; HF data communications; bandwidths of various modes

G8B01 (A)
What receiver stage combines a 14.250 MHz input signal with a 13.795 MHz
oscillator signal to produce a 455 kHz intermediate frequency (IF) signal?
A. Mixer
B. --------
C. --------
D. --------
~~

G8B02 (B)
If a receiver mixes a 13.800 MHz VFO with a 14.255 MHz received signal to produce
a 455 kHz intermediate frequency (IF) signal, what type of interference will a
13.345 MHz signal produce in the receiver?
A. --------
B. Image response
C. --------
D. --------
~~

G8B03 (A)
What stage in a transmitter would change a 5.3 MHz input signal to 14.3 MHz?
A. A mixer
B. --------
C. --------
D. --------
~~

G8B04 (D)
What is the name of the stage in a VHF FM transmitter that selects a harmonic of an HF signal to reach the desired operating frequency?
A. --------
B. --------
C. --------
D. Multiplier
~~

G8B05 (C)
Why isn't frequency modulated (FM) phone used below 29.5 MHz?
A. --------
B. --------
C. The bandwidth would exceed FCC limits
D. --------
~~

G8B06 (D)
What is the total bandwidth of an FM-phone transmission having a 5 kHz
deviation and a 3 kHz modulating frequency?
A. --------
B. --------
C. --------
D. 16 kHz
~~

G8B07 (B)
What is the frequency deviation for a 12.21-MHz reactance-modulated oscillator in a 5-kHz deviation, 146.52-MHz FM-phone transmitter?
A. --------
B. 416.7 Hz
C. --------
D. --------
~~

G8B08 (C)
How is frequency shift related to keying speed in an FSK signal?
A. --------
B. --------
C. Greater keying speeds require greater frequency shifts
D. --------
~~

# SIMPLIFIED STUDY GUIDE FOR GENRAL CLASS...

G8B09 (B)
What do RTTY, Morse code, PSK31 and packet communications have in common?
A. --------
B. They are digital modes
C. --------
D. --------
~~

G8B10 (B)
When transmitting a data mode signal, why is it important to know the duty cycle of the mode you are using?
A. --------
B. To prevent damage to your transmitter's final output stage
C. --------
D. --------
~~

G8B11 (D)
What part of the 20 meter band is most commonly used for PSK31 operation?
A. --------
B. --------
C. --------
D. Below the RTTY segment, near 14.070 MHz
~~

G8B12 (A)
What is another term for the mixing of two RF signals?
A. Heterodyning
B. --------
C. --------
D. --------
~~

# 12

# ANTENNAS

SUBELEMENT G9 – ANTENNAS AND FEEDLINES [4 exam questions – 4 groups]

G9A - Antenna feedlines: characteristic impedance, and attenuation; SWR calculation, measurement and effects; matching networks

G9A01 (A)
Which of the following factors help determine the characteristic impedance of a parallel conductor antenna feedline?
A. The distance between the centers of the conductors and the radius of the
   conductors
B. --------
C. --------
D. --------
~~

G9A02 (B)
What is the typical characteristic impedance of coaxial cables used for antenna feedlines at amateur stations?
A. --------
B. 50 and 75 ohms
C. --------
D. --------
~~

G9A03 (D)
What is the characteristic impedance of flat ribbon TV type twin lead?
A. --------
B. --------
C. --------
D. 300 ohms

~~

G9A04 (C)
What is a common reason for the occurrence of reflected power at the point where a feedline connects to an antenna?
A. --------
B. --------
C. A difference between feedline impedance and antenna feed point impedance
D. --------

~~

G9A05 (D)
What must be done to prevent standing waves on an antenna feedline?
A. --------
B. --------
C. --------
D. The antenna feed point impedance must be matched to the characteristic impedance
   of the feedline

~~

G9A06 (C)
Which of the following is a reason for using an inductively coupled matching network between the transmitter and parallel conductor feed line feeding an antenna?
A. --------
B. --------
C. To match the unbalanced transmitter output to the balanced parallel conductor
   feedline
D. --------

~~

G9A07 (B)
How does the attenuation of coaxial cable change as the frequency of the signal it is carrying increases?
A. --------
B. It increases
C. --------
D. --------
~~

G9A08 (D)
In what values are RF feed line losses usually expressed?
A. --------
B. --------
C. --------
D. dB per 100 ft
~~

G9A09 (A)
What standing-wave-ratio will result from the connection of a 50-ohm feed line to a non-reactive load having a 200-ohm impedance?
A. 4:1
B. --------
C. --------
D. --------
~~

G9A10 (D)
What standing-wave-ratio will result from the connection of a 50-ohm feed line to a non-reactive load having a 10-ohm impedance?
A. --------
B. --------
C. --------
D. 5:1
~~

G9A11 (B)
What standing-wave-ratio will result from the connection of a 50-ohm feed line to a non-reactive load having a 50-ohm impedance?
A. --------
B. 1:1
C. --------
D. --------
~~

# SIMPLIFIED STUDY GUIDE FOR GENRAL CLASS...

G9A12 (A)
What would be the SWR if you feed a vertical antenna that has a 25-ohm feed-point impedance with 50-ohm coaxial cable?
A. 2:1
B. --------
C. --------
D. --------
~~

G9A13 (C)
What would be the SWR if you feed a folded dipole antenna that has a 300-ohm feed-point impedance with 50-ohm coaxial cable?
A. --------
B. --------
C. 6:1
D. --------
~~

G9A14 (B)
If the SWR on an antenna feedline is 5 to 1, and a matching network at the transmitter end of the feedline is adjusted to 1 to 1 SWR, what is the resulting SWR on the feedline?
A. --------
B. 5 to 1
C. --------
D. --------
~~


G9B - Basic antennas


G9B01 (B)
What is one disadvantage of a directly fed random-wire antenna?
A. --------
B. You may experience RF burns when touching metal objects in your station
C. --------
D. --------
~~

G9B02 (D)
What is an advantage of downward sloping radials on a ground-plane antenna?
A. --------
B. --------
C. --------
D. They can be adjusted to bring the feed-point impedance closer to 50 ohms
~~

G9B03 (B)
What happens to the feed-point impedance of a ground-plane antenna when its radials are changed from horizontal to downward-sloping?
A. --------
B. It increases
C. --------
D. --------
~~

G9B04 (A)
What is the low angle azimuthal radiation pattern of an ideal half-wavelength dipole antenna installed 1/2 wavelength high and parallel to the earth?
A. It is a figure-eight at right angles to the antenna
B. --------
C. --------
D. --------
~~

G9B05 (C)
How does antenna height affect the horizontal (azimuthal) radiation pattern of a horizontal dipole HF antenna?
A. --------
B. --------
C. If the antenna is less than 1/2 wavelength high, the azimuthal pattern is almost omnidirectional
D. --------
~~

G9B06 (C)
Where should the radial wires of a ground-mounted vertical antenna system be placed?
A. --------
B. --------
C. On the surface or buried a few inches below the ground
D. --------
~~

G9B07 (B)
How does the feed-point impedance of a 1/2 wave dipole antenna change as the antenna is lowered from 1/4 wave above ground?
A. --------
B. It steadily decreases
C. --------
D. --------
~~

G9B08 (A)
How does the feed-point impedance of a 1/2 wave dipole change as the feed-point location is moved from the center toward the ends?
A. It steadily increases
B. --------
C. --------
D. --------
~~

G9B09 (A)
Which of the following is an advantage of a horizontally polarized as compared to vertically polarized HF antenna?
A. Lower ground reflection losses
B. --------
C. --------
D. --------
~~

G9B10 (D)
What is the approximate length for a 1/2-wave dipole antenna cut for 14.250 MHz?
A. --------
B. --------
C. --------
D. 32.8 feet
~~

G9B11 (C)
What is the approximate length for a 1/2-wave dipole antenna cut for 3.550 MHz?
A. --------
B. --------
C. 131.8 feet
D. --------

G9B12 (A)
What is the approximate length for a 1/4-wave vertical antenna cut for 28.5 MHz?
A. 8.2 feet
B. --------
C. --------
D. --------

G9C - Directional antennas

G9C01 (A)
How can the SWR bandwidth of a Yagi antenna be increased?
A. Use larger diameter elements
B. --------
C. --------
D. --------

G9C02 (B)
What is the approximate length of the driven element of a Yagi antenna?
A. --------
B. 1/2 wavelength
C. --------
D. --------

G9C03 (B)
Which statement about a three-element single-band Yagi antenna is true?
A. --------
B. The director is normally the shortest parasitic element
C. --------
D. --------

G9C04 (A)
Which statement about a Yagi antenna is true?
A. The reflector is normally the longest parasitic element
B. --------
C. --------
D. --------

G9C05 (A)
What is one effect of increasing the boom length and adding directors to a Yagi antenna?
A. Gain increases
B. --------
C. --------
D. --------
~~

G9C06 (C)
Which of the following is a reason why a Yagi antenna is often used for radio communications on the 20 meter band?
A. --------
B. --------
C. It helps reduce interference from other stations to the side or behind the antenna
D. --------
~~

G9C07 (C)
What does "front-to-back ratio" mean in reference to a Yagi antenna?
A. --------
B. --------
C. The power radiated in the major radiation lobe compared to the power
   radiated in exactly the opposite direction
D. --------
~~

G9C08 (D)
What is meant by the "main lobe" of a directive antenna?
A. --------
B. --------
C. --------
D. The direction of maximum radiated field strength from the antenna
~~

G9C09 (A)
What is the approximate maximum theoretical forward gain of a 3 Element Yagi antenna?
A. 9.7 dBi
B. --------
C. --------
D. --------

G9C10 (D)
Which of the following is a Yagi antenna design variable that could be adjusted to optimize forward gain, front-to-back ratio, or SWR bandwidth?
A. --------
B. --------
C. --------
D. All of these choices are correct

G9C11 (A)
What is the purpose of a "gamma match" used with Yagi antennas?
A. To match the relatively low feed-point impedance to 50 ohms
B. --------
C. --------
D. --------

G9C12 (D)
Which of the following describes a common method for insulating the driven element of a Yagi antenna from the metal boom when using a gamma match?
A. --------
B. --------
C. --------
D. None of these answers are correct. No insulation is needed

G9C13 (A)
Approximately how long is each side of a cubical-quad antenna driven element?
A. 1/4 wavelength
B. --------
C. --------
D. --------

G9C14 (B)
How does the forward gain of a 2-element cubical-quad antenna compare to the forward gain of a 3 element Yagi antenna?
A. --------
B. About the same
C. --------
D. --------

G9C15 (B)
Approximately how long is each side of a cubical-quad antenna reflector
element?
A. --------
B. Slightly more than 1/4 wavelength
C. --------
D. --------
~~

G9C16 (D)
How does the gain of a two element delta-loop beam compare to the gain of a two element cubical quad antenna?
A. --------
B. --------
C. --------
D. About the same
~~

G9C17 (B)
Approximately how long is each leg of a symmetrical delta-loop antenna
Driven element?
A. --------
B. 1/3 wavelengths
C. --------
D. --------
~~

G9C18 (D)
Which of the following antenna types consists of a driven element and some combination of parasitically excited reflector and/or director elements?
A. --------
B. --------
C. --------
D. A Yagi antenna
~~

G9C19 (C)
What type of directional antenna is typically constructed from 2 square loops of wire each having a circumference of approximately one wavelength at the operating frequency and separated by approximately 0.2 wavelength?
A. --------
B. --------
C. A cubical quad antenna
D. --------
~~

G9C20 (A)
What happens when the feed-point of a cubical quad antenna is changed from the center of the lowest horizontal wire to the center of one of the vertical wires?
A. The polarization of the radiated signal changes from horizontal to vertical
B. --------
C. --------
D. --------
~~

G9C21 (D)
What configuration of the loops of a cubical-quad antenna must be used for the antenna to operate as a beam antenna, assuming one of the elements is used as a reflector?
A. --------
B. --------
C. --------
D. The reflector element must be approximately 5% longer than the driven element
~~


G9D - Specialized antennas


G9D01 (D)
What does the term "NVIS" mean as related to antennas?
A. --------
B. --------
C. --------
D. Near Vertical Incidence Skywave
~~

# SIMPLIFIED STUDY GUIDE FOR GENRAL CLASS...

G9D02 (B)
Which of the following is an advantage of an NVIS antenna?
A. --------
B. High vertical angle radiation for short skip during the day
C. --------
D. --------
~~

G9D03 (D)
At what height above ground is an NVIS antenna typically installed?
A. --------
B. --------
C. --------
D. Between 1/10 and 1/4 wavelength
~~

G9D04 (B)
How does the gain of two 3-element horizontally polarized Yagi antennas spaced vertically 1/2 wave apart from each other typically compare to the gain of a single 3-element Yagi?
A. --------
B. Approximately 3 dB higher
C. --------
D. --------
~~

G9D05 (D)
What is the advantage of vertical stacking of horizontally polarized Yagi antennas?
A. --------
B. --------
C. --------
D. Narrows the main lobe in elevation
~~

G9D06 (A)
Which of the following is an advantage of a log periodic antenna?
A. Wide bandwidth
B. --------
C. --------
D. --------
~~

G9D07 (A)
Which of the following describes a log periodic antenna?
A. Length and spacing of the elements increases logarithmically from one end of
   the boom to the other
B. --------
C. --------
D. --------
~~

G9D08 (B)
Why is a Beverage antenna generally not used for transmitting?
A. --------
B. It has high losses compared to other types of antennas
C. --------
D. --------
~~

G9D09 (B)
Which of the following is an application for a Beverage antenna?
A. --------
B. Directional receiving for low HF bands
C. --------
D. --------
~~

G9D10 (D)
Which of the following describes a Beverage antenna?
A. --------
B. --------
C. --------
D. A very long and low receiving antenna that is highly directional
~~

# SIMPLIFIED STUDY GUIDE FOR GENRAL CLASS...

G9D11 (D)
Which of the following is a disadvantage of multiband antennas?
A. --------
B. --------
C. --------
D. They have poor harmonic rejection
~~

G9D12 (A)
What is the primary purpose of traps installed in antennas?
A. To permit multiband operation
B. --------
C. --------
D. --------
~~

# 13

# SAFETY

SUBELEMENT G0 – ELECTRICAL AND RF SAFETY [2 Exam Questions – 2 groups]

G0A - RF safety principles, rules and guidelines; routine station evaluation

G0A01 (A)
What is one way that RF energy can affect human body tissue?
A. It heats body tissue
B. --------
C. --------
D. --------
~~

G0A02 (B)
Which property is NOT important in estimating if an RF signal exceeds the maximum permissible exposure (MPE)?
A. --------
B. Its critical angle
C. --------
D. --------
~~

# SIMPLIFIED STUDY GUIDE FOR GENRAL CLASS…

G0A03 (B)
Which of the following has the most direct effect on the permitted exposure level of RF radiation?
A. --------
B. The power level and frequency of the energy
C. --------
D. --------
~~

G0A04 (D)
What does "time averaging" mean in reference to RF radiation exposure?
A. --------
B. --------
C. --------
D. The total RF exposure averaged over a certain time
~~

G0A05 (A)
What must you do if an evaluation of your station shows RF energy radiated from your station exceeds permissible limits?
A. Take action to prevent human exposure to the excessive RF fields
B. --------
C. --------
D. --------
~~

G0A06 (C)
Which transmitter(s) at a multiple user site is/are responsible for RF safety compliance?
A. --------
B. --------
C. Any transmitter that contributes 5% or more of the MPE
D. --------
~~

G0A07 (A)
What effect does transmitter duty cycle have when evaluating RF exposure?
A. A lower transmitter duty cycle permits greater short-term exposure levels
B. --------
C. --------
D. --------

G0A08 (C)
Which of the following steps must an amateur operator take to ensure compliance with RF safety regulations?
A. --------
B. --------
C. Perform a routine RF exposure evaluation
D. --------

G0A09 (B)
What type of instrument can be used to accurately measure an RF field?
A. --------
B. A calibrated field-strength meter with a calibrated antenna
C. --------
D. --------

G0A10 (D)
What do the RF safety rules require when the maximum power output capability of an otherwise compliant station is reduced?
A. --------
B. --------
C. --------
D. No further action is required

G0A11 (C)
What precaution should you take if you install an indoor transmitting antenna?
A. --------
B. --------
C. Make sure that MPE limits are not exceeded in occupied areas
D. --------

G0A12 (B)
What precaution should you take whenever you make adjustments or repairs to an antenna?
A. --------
B. Turn off the transmitter and disconnect the feedline
C. --------
D. --------

# SIMPLIFIED STUDY GUIDE FOR GENRAL CLASS…

~~

G0A13 (D)
What precaution should be taken when installing a ground-mounted antenna?
A. --------
B. --------
C. --------
D. It should be installed so no one can be exposed to RF radiation in excess of
   maximum permissible limits

~~

G0A14 (D)
What is one thing that can be done if evaluation shows that a neighbor might receive more than the allowable limit of RF exposure from the main lobe of a directional antenna?
A. --------
B. --------
C. --------
D. Take precautions to ensure that the antenna cannot be pointed at their house

~~

G0A15 (D) [97.13(c)(1)]
How can you determine that your station complies with FCC RF exposure regulations?
A. By calculation based on FCC OET Bulletin 65
B. By calculation based on computer modeling
C. By measurement of field strength using calibrated equipment
D. All of these choices are correct

~~

G0B - Safety in the ham shack: electrical shock and treatment, grounding, fusing, interlocks, wiring, antenna and tower safety

G0B01 (A)
Which wire(s) in a four-conductor line cord should be attached to fuses or circuit breakers in a device operated from a 240-VAC single-phase source?
A. Only the "hot" (black and red) wires
B. --------
C. --------
D. --------
~~

G0B02 (C)
What is the minimum wire size that may be safely used for a circuit that draws up to 20 amperes of continuous current?
A. --------
B. --------
C. AWG number 12
D. --------
~~

G0B03 (D)
Which size of fuse or circuit breaker would be appropriate to use with a circuit that uses AWG number 14 wiring?
A. --------
B. --------
C. --------
D. 15 amperes
~~

G0B04 (A)
What is the mechanism by which electrical shock can be lethal?
A. Current through the heart can cause the heart to stop pumping
B. --------
C. --------
D. --------
~~

# SIMPLIFIED STUDY GUIDE FOR GENRAL CLASS...

G0B05 (B)
Which of the following conditions will cause a Ground Fault Circuit Interrupter (GFCI) to disconnect the 120 or 240 Volt AC line power to a device?
A. --------
B. Current flowing from the hot wire to ground
C. --------
D. --------
~~

G0B06 (D)
Why must the metal chassis of every item of station equipment be grounded (assuming the item has such a chassis)?
A. --------
B. --------
C. --------
D. It ensures that hazardous voltages cannot appear on the chassis
~~

G0B07 (B)
Which of the following should be observed for safety when climbing on a tower using a safety belt or harness?
A. --------
B. Always attach the belt safety hook to the belt "D" ring with the hook opening away
   from the tower
C. --------
D. --------
~~

G0B08 (B)
What should be done by any person preparing to climb a tower that supports electrically powered devices?
A. --------
B. Make sure all circuits that supply power to the tower are locked out and tagged
C. --------
D. --------
~~

G0B09 (D)
Why is it not safe to use soldered joints with the wires that connect the base of a tower to a system of ground rods?
A. --------
B. --------
C. --------
D. A soldered joint will likely be destroyed by the heat of a lightning strike
~~

G0B10 (A)
Which of the following is a danger from lead-tin solder?
A. Lead can contaminate food if hands are not washed carefully after handling
B. --------
C. --------
D. --------
~~

G0B11 (D)
Which of the following is good engineering practice for lightning protection grounds?
A. --------
B. --------
C. --------
D. They must be bonded together with all other grounds
~~

G0B12 (C)
What is the purpose of a transmitter power supply interlock?
A. --------
B. --------
C. To ensure that dangerous voltages are removed if the cabinet is opened
D. --------
~~

G0B13 (B)
Which of the following is the most hazardous type of electrical energy?
A. --------
B. 60 cycle Alternating current
C. --------
D. --------
~~

G0B14 (B)

What is the maximum amount of electrical current flow through the human body that can be tolerated safely?
A. --------
B. 50 microamperes
C. --------
D. --------
~~

**GOOD LUCK ON THE TEST! Go through the question pool again if this is your first time through. You won't regret it!**

END OF BOOK

LaVergne, TN USA
21 July 2010
190365LV00008B/156/P